The Politics of Baby Foods

GLOBAL POLITICS

A Series edited by Peter Willetts,
The City University, London

Already published

THE NON-ALIGNED IN HAVANA
Documents of the Sixth Summit Conference and an Analysis of their Significance for the Global Political System
Peter Willetts

PRESSURE GROUPS IN THE GLOBAL SYSTEM
The Transnational Relations of Issue-Orientated Non-Governmental Organizations
Edited by Peter Willetts

AN INTERNATIONAL LAW OF GUERRILLA WARFARE
The Global Politics of Law-Making
Keith Suter

First published in Great Britain in 1986 by
Frances Pinter (Publishers) Limited
25 Floral Street, London WC2E 9DS

British Library Cataloguing in Publication Data

Chetley, Andrew
The politics of babyfoods: successful challenges to an international marketing strategy.
1. Baby foods industry 2. Pressure groups
I. Title
664 ′.62 ′0688 HD9970.5.IS42
ISBN 0-86187-293-2

Typeset by Joshua Associates Limited, Oxford
Printed by Biddles of Guildford Ltd

The Politics of Baby Foods

Successful challenges to an international marketing strategy

Andrew Chetley

Frances Pinter (Publishers), London

Contents

For Rachel and Rebecca

Series Editor's preface

Andy Chetley has written a very unusual book on international relations. It is an insider's report on the campaign by pressure groups to restrict the marketing of dried milk as a baby food and to limit the irresponsible operations of some transnational companies. The use of dried milk as a replacement for breast feeding has been responsible for the deaths of, literally, countless numbers of babies, measured not in thousands but in millions of deaths. The companies selling dried milk have argued that it is only incorrect mixing of the powder or use of unclean water that is a risk to health. The pressure groups have argued that in poor countries, without publicly-supplied clean water and with low levels of literacy and health education, dried milk will inevitably be a dangerous product and that, even in the best of conditions, it is not a complete substitute for breast milk. The issue has been of great controversy, eventually involving all the world's governments in the World Health Organization's decision to establish an International Code of Marketing of Breast-milk Substitutes. The campaign was so effective that the Code was adopted virtually unanimously, by 118 votes to 1, with the sole opposing vote coming from the United States' government.

Although this is the story of a conflict in which a 'superpower', as one of the major protagonists, suffered a substantial blow to its prestige, it is not being told in the memoirs of some leading diplomat or foreign minister. Andy Chetley is an ordinary person whose commitment to the issue gradually turned him into an expert campaigner. He has become one of the key figures, working in a diverse group of non-governmental organizations strung around the globe, who placed this issue on the public agenda. Since the early 1970s, he has been active, first within the British development agency, War on Want, and from May 1984 as the General Administrator of the International Baby Foods Action Network (IBFAN). Andy has also attended all the crucial World Health Organization meetings on the subject; so most of the events are being recounted by someone who has been a participant in them. However, despite his commitment, this book is not a one-sided account from a limited perspective, but the full story of the campaign. A remarkable range of sources has been used, not only describing the work of all the other medical, church, consumer, development and women's groups which eventually came together in IBFAN, but also including the view from the other side of the battle, in the words of the leaders of the transnational companies and even extracts from top-secret United States' diplomatic telegrams. The most important of the unpublished sources for this book are now available, for those who wish to pursue some aspects of the story in more detail, from the *Global Village Meetings Newspapers Archive* in the library of The City University, London.

The aim of the Global Politics Series of books is to move away from the idea that the study of international relations is simply concerned with the diplomatic and military relations between coherent units called states. More attention should be given to non-governmental organizations (NGOs) and intergovernmental organizations, such as the United Nations or the World Health Organization (WHO), as participants in global politics. More attention should be given to economic and social issues, as well as to the diversity among what have been traditionally viewed as 'political issues'. We must change our approach, from one based on the idea that politics involves a struggle for 'power', to seeing politics as contention over issues.

I invited Andy Chetley to write about the baby foods campaign, because it would clearly be an account of the creation, maintenance and expansion of an issue. It would also be an example of an issue where governments were not the sole actors: indeed, they were not even the main actors. By the end of 1979, the 'baby foods Cold War' was a bipolar conflict between two global alliances, the International Council of Infant Food Industries (ICIFI) and the International Baby Food Action Network (IBFAN). Governments were drawn in and forced to take sides in this bipolar Cold War between two NGOs and one of the main arenas for the conflict was an intergovernmental organization, the World Health Organization. The issue can only be understood by considering the interactions of all three types of actors: NGOs, governments and intergovernmental organizations.

Traditional writers, who believe that the East/West Cold War shows the 'real' nature of the international system, like to distinguish the 'high politics' of security, diplomacy and national prestige from the 'low politics' of economic and social issues. The fact that governments are not dominant on baby foods can be treated as peripheral to the traditionalists' world view by labelling the issue as low politics. Describing other issues as high politics appears to carry four implications: that the issue is of high importance; that it will take priority in people's time and concern over other issues; that it will be dealt with by high-level decision-makers; and that the more 'powerful' governments will determine outcomes.

No doubt some might regard the baby foods issue as unimportant. It would be difficult to find an issue more removed from the nuclear arms race and East/West relations than how babies are fed. It concerns women and children and is not of significance to 'real' men. Yet, in terms of the danger to human life, which is one of the standards used to assert the importance of East/West relations, baby foods are also important. Misuse of dried milk has almost certainly resulted in a similar number of deaths to those killed in all the wars since 1945.[1]

To say that 'high politics' issues command the attention of decision-makers begs the question of which decision-makers we consider to be worthy of attention. If only governments are worthy of analysis, then it is true that the pressure groups have had to force the marketing of baby foods to their attention and hardly any cabinets will have considered the issue more than once or twice in the last fifteen years. But governments have very many issues on their agenda and cabinets seldom consider any issue on more than a few occasions. It is individual ministries that do all the detailed decision-making and that is true for defence as much as it is for health issues. Within the health ministries and development agencies of the world and in the delegations to WHO, there are government officials who have had to make baby foods their top priority in recent years. This book is tribute to the fact that, for some pressure group activists and some transnational company executives, the issue has been top priority for more than a decade.

We do not know the details of exactly how the critical decision of whether to vote for the WHO Code was taken by each government. However, we do know that it was not a routine 'low politics' decision by junior health officials. In the United States, Andy recounts the full process of Senate hearings, conflict between departments, appeal to the White House and defeats for the president in Congress. Similarly in Britain (though this is not in the book), MPs lobbied ministers, tabled parliamentary questions, signed Early Day Motions and held an adjournment debate. In another twenty-one countries there might not have been the same controversy, but high-level decision-makers must have been involved in the preparation of legislation to implement the WHO Code.

Finally, the importance of an issue-based approach rather than a power-based approach is shown most clearly in the total isolation of the United States' government. For those who favoured a vote against the WHO Code it was fundamental to the American 'national interest' to maintain an untrammelled, free-market international economic system and to prevent international organizations, such as WHO, developing the authority to implement global regulation of transnational companies. Both these points were central to the Reagan administration's political identity and for that reason the decision to vote 'No' was not surprising. But how could the most 'powerful' government in the world be in a minority of one? This outcome cannot possibly be explained in power terms, whereas it is explicable in terms of the ideological gulf between the United States' government and all other governments on this specific issue.

Economic and social issues do not belong to some secondary realm of 'low politics'. Baby feeding matters a great deal, it has been a high-priority issue for many people and has involved top-level government officials and politicians. There has also been progress derived solely from the technical and moral strengths of the arguments put by ordinary people working against entrenched economic and political interests.

This book strongly complements two previous volumes in the Global Politics Series. *Pressure Groups in the Global System* offers a general study of NGOs to provide a broader context for the baby foods issue, and *An International Law of Guerrilla Warfare* offers a comparison of how NGOs similarly forced another issue on to the agendas of governments.

Andy Chetley says this is 'a book about ordinary people and about the way people organize for change' and that 'the organization of the campaign provides a new model for cooperation on important social, economic and political issues'.

Part of his motivation for telling the story is that the lessons of the campaign should be available to those who want to work on other issues. Already another network like IBFAN, Health Action International, is covering the marketing of pharmaceuticals, and the issue of the export of dangerous products, such as pesticides, is attracting attention among NGOs. The book is in the Global Politics Series because it helps to provide a second, different type of challenge to the existing order. It demonstrates that traditional modes of thinking about international relations cannot handle the complexities of modern world politics.

Peter Willetts
The City University, London
August 1985

Note

1. The claim that the misuse of dried baby milk has led to tens of millions of deaths might seem to be exaggerated, but from *The State of the World's Children 1985* (London, Oxford University Press for UNICEF, 1984) one can make an estimate of the order of magnitude. A number of research studies have shown that bottle-fed babies are significantly more likely to die than breast-fed babies (see p. 87). Taking a conservative assumption that the danger is twice as high, the proportion of infant deaths due to bottle feeding would be $b/(1+b)$, where b is the proportion which were bottle-fed (see table at the end of this note). The figures in the first three columns of the table below are taken from the Statistics section of UNICEF's 1985 volume. A second conservative feature of the estimate is the omission of the thirty-one countries in the low IMR category. The last three columns are not very reliable estimates because the figures for the percentage of babies which are bottle fed are not very reliable (derived from pp. 114–15, the percentage that do not bottle feed for at least three months). Nevertheless, the figure of 1 million deaths in 1982, due to misuse of dried milk in bottle feeding, must be the correct order of magnitude. As war has claimed somewhere between 16 and 30 million civilian and military deaths since 1945 (depending on which estimates one takes), the estimated total for bottle-feeding deaths in the thirty–forty years in which the use of dried milk has become common must be within a similar range.

World infant mortality rates (IMR)

	No. of countries	Births*	IMR/ 1,000	Infant deaths*	% on bottle	Prop. of deaths from bottle	Deaths due to bottle*
Very high IMR	43	53,937	140	7,551	6	6/106	427
High IMR	29	24,419	90	2,198	13	13/113	253
Middle IMR	27	31,198	38	1,186	37	37/137	320
Total	99	109,554	–	10,935	–	–	1,000

* 1982 data in thousands.

Acknowledgements

The list of people who deserve thanks for making this book possible—both because of the advice, ideas, assistance and information they gave, and because of the efforts they have made to improve infant feeding practices—would of itself nearly fill the pages of a book. It includes health workers in many different parts of the world, UN officials in several organizations, policy-makers in different governments, development workers, community leaders, members of pressure groups, journalists, mothers and parents who have helped make this a book about ordinary people and about the way people organize for change.

About a dozen people working with the International Baby Food Action Network at various times provided the inspiration, encouragement and cajoling necessary to ensure that the book was finally completed.

Special thanks goes to Thierry Lemaresquier whose early thoughts about the concepts in the book and whose continual encouragement helped immeasurably. John Noyce deserves the credit for excellent library research which has helped to make this a more complete work. Peter Willetts' patient and professional editing has been essential in keeping me on the right track.

Note on terminology

The major groups or 'actors' described in this book fall into different categories, none of which is particularly adequately described in existing terminology. However, short of inventing some new and unwieldy terms, I have resorted to using the following nomenclature and offer, by way of both apology and definition, the following explanations.

Developing countries: Many terms are used in literature to describe those nations in Asia, Africa and Latin America whose gross national products and other economic and social indices are less favourable than, for example, the countries of Western Europe. Collectively, they are called the *Third World*, or *underdeveloped* nations, the *poor world*, the *South*, *industrializing* nations, or *less developed countries (LDCs)*. Throughout this book, 'developing' has been used (except where other terms are found in direct quotes from other sources), even though this begs the question about whether these countries are actually developing and also about the so-called 'developed' countries, for it can be argued that all nations are, or should be, developing.

Industrialized countries: This term broadly covers countries in Western and Eastern Europe, North America, Japan, Australia and New Zealand. These countries are sometimes referred to as the *North*, the *West* (with the exception of Eastern Europe), *developed* nations. Readers should, however, bear in mind that the term used does not imply homogeneity either among or within these states, as they are at different stages of industrialization when compared to each other, and have different levels of industrialization internally.

Transnational corporations (TNCs): This term refers to those commercial enterprises that operate across national borders. It generally refers to the world's largest manufacturing and trading concerns. Other terms that have been used to describe them include *multinational corporations (MNCs), international firms/companies/enterprises*.

Non-governmental organizations (NGOs): At the risk of arousing the ire of purists, who maintan that the term NGO should only apply to those organizations that receive *official* recognition from the United Nations system, the European Economic Commission, or similar intergovernmental bodies, I have, for simplicity's sake, used the term to describe a whole host of groups and organizations: *development agencies, consumer groups, specific interest groups, development education groups, pressure groups, public interest advocates*, and *coalitions* or *networks* of any of the groups. A summary definition of an NGO, as used in this book, is an organization that is neither government, intergovernmental, nor profit-motivated. A preferred term would be *people's movements*, as this would imply the concept of popular support, and the ideas of activity, change and growth.

Infant formula: Although a very specific product, the term has been used in a more generic sense throughout the book to include products which are *used* to feed infants, even though they may not have been designed specifically to do so. The term is used in preference to *breast milk substitutes* in order to avoid the possible assumption that there is an interchangeable and complete equivalent substitute for breast milk. Other terms that may occur include *powdered baby milks, infant milks/foods*.

Abbreviations

AgDW	Arbeitsgruppe Dritte Welt (Third World Working Group—Switzerland)
BMAC	Baby Milk Action Coalition (UK)
CCPD	Commission on the Churches' Participation in Development (of the WCC)
Codex	WHO/FAO Codex Alimentarius Commission
CTC	Centre on Transnational Corporations (of the UN)
DHSS	Department of Health and Social Security (UK)
ECOSOC	Economic and Social Council (of the UN)
EEC	European Economic Commission/Community
EPPC	Ethics and Public Policy Centre (USA)
FAO	Food and Agriculture Organization (of the UN)
FDA	Food and Drug Administration (USA)
FMF	Food Manufacturers' Federation (UK)
GIFA	Geneva Infant Feeding Association (Switzerland)
HAI	Health Action International
IBFAN	International Baby Food Action Network
ICCR	Interfaith Centre on Corporate Responsibility (USA)
ICIFI	International Council of Infant Food Industries
IDACE	Association of Dietetic Food Industries of the EEC
IFDA	International Foundation for Development Alternatives
INBC	International Nestlé Boycott Committee
INFACT	Infant Formula Action Coalition (USA)
IOCU	International Organization of Consumers' Unions
IPA	International Paediatrics Association
ISIS	Women's International Cross-Cultural Exchange
NCC	National Council of Churches (USA)
NGLS	Non-Governmental Liaison Service (of the UN)
NGO	non-governmental organization
NIFAC	Nestlé Infant Formula Audit Commission
OECD	Organization for Economic Cooperation and Development
PAG	Protein-Calorie Advisory Group (of the UN)
R&D	research and development
SAFEP	Swiss Information Groups for Development Policy
SCM	sweetened condensed milk
SMP	skimmed milk powder
TNC	transnational corporation
UNCTAD	United Nations Conference on Trade and Development

UNICEF	United Nations Children's Fund
US AID	United States Agency for International Development
WCC	World Council of Churches
WHA	World Health Assembly
WHO	World Health Organization

Introduction

> Power is exercised through the state and the market, but people have an autonomous power, legitimately theirs, which they are increasingly asserting.
>
> 'Building Blocks for Alternative Development Strategies',
> *IFDA Dossier*, **17**, May/June 1980

The sovereign state is normally identified as the individual unit of power in the world. During the past few centuries of history, events have been expressed in terms of the actions of nations. Exploration, colonization, wars, independence, treaties and agreements—all are described as the preserve of separate geographical 'units'.

Hidden behind the facade of the state, however, is the driving force that first established and, later, sustained the growth of nations—the economic power of commerce. As the Industrial Revolution swept through Northern Europe, it brought with it the demand for clearly defined and controllable markets, which included political control. The city states and local power preserves of the aristocracies were too fragmented to allow for the easy passage of new products. As Toffler puts it,

> How could businessmen buy and sell over a large territory if, outside their own community, they ran into a maze of different duties, taxes, labour regulations and currencies? For the new technologies to pay off, local economies had to be consolidated into a single national economy. This meant a national division of labour and a national market for commodities and capital. All this, in turn, required national political consolidation as well.[1]

That economic imperative slowly infused the political and social systems of most of the now-industrialized nations (including, with some local variations, the nations of Eastern Europe). Structures and attitudes that were essential for protecting and increasing the growth of the market economy became woven into every fabric of daily life. Schools, churches, the work-place, even the design of the communities in which people lived, exuded a constant conditioning, urging all members of society to take pride in the nation, work for it, fight for it and, if necessary, die for it. Above all, it was important not to examine or question too closely.

Progress was the important catchword, and only a few eccentric commentators dared to ask 'at what price?'. The few dissenters provided a safety valve, a release of pressure. Other safety valves became incorporated in the concept of democracy, with the illusion of control and accountability provided through a system of voting for representatives every few years.

Those who controlled the economic forces at play in society could not afford to have the whims of different political perspectives interfering with economic growth however, and they worked hard to entangle economic interests with national interests to such an extent that the two could not be easily separated. Charles Wilson, president of General Motors, summed up the situation succinctly when he said, 'I have always thought that what was good for the country was equally good for General Motors—and vice versa'.[2] But if these activities were good for one nation, and if a national division of labour and a national market-place were beneficial, how much more beneficial would an international market be? The commodities, the products, the techniques, the attitudes and the conditioning all began to be exported. The world slowly became a global market-place.

If, to start with, all the parties in the market-place had had an equal footing, there could be some real advantages to the streamlining, standardization and concentration that have occurred. However, most of the world started late—a direct result of the colonization process operated by the European nations. It was a process that guaranteed cheap resources, cheap labour and captive markets for their industries. Eduardo Galeano describes the process:

> The division of labour among nations is that some specialise in winning, and others in losing. Our part of the world, known today as Latin America, was precocious: it has specialised in losing ever since those remote times when the Renaissance Europeans ventured across the ocean and buried their teeth in the throats of the Indian civilisations. Centuries passed, and Latin America perfected its role . . . our region still works as a menial. It continues to exist at the service of others' needs.[3]

The overall result, viewed from almost any perspective is disastrous. Approximately one-fifth of the world's population live in severe destitution, described by the president of the World Bank as a condition 'below any rational definition of human decency'.[4] About half the world's population lack easy access to clean water. Nearly 400 million people throughout the world are un- or under-employed. A United Nations publication sums up the global picture:

> Existing systems of production and exchange do not seem to work in the interests of people or of their natural environment. Billions of dollars are spent yearly on armaments while essential social services are ignored for lack of funds. In many countries there is mass hunger while in others there is a tremendous waste of food. Within countries and among them there are huge gaps between the living standards of the rich and those of the poor. Levels of income, technology, production and trade are all widely different, with some having too little and others so much that their waste destroys the natural environment.[5]

In short, the document concludes that 'what we have today is a world economy that is out of control'.

Clearly, no single state can either assume total responsibility for this crisis or overcome it. This is due not only to the magnitude of the crisis but also to the presence of another set of powerful actors on the international stage—transnational corporations (TNCs). According to the Brandt Report, TNCs control between one-quarter and one-third of all world production with about 30 per cent of all world trade going on between parent firms and their affiliates or subsidiaries. By 1976, total sales of the foreign affiliates of TNCs were estimated at $836 billion, approximately equal to the gross national product of all developing countries with the exception of the oil exporters.[6]

Although international trade has flourished for hundreds of years, the growth of the supercompanies or TNCs is a recent phenomenon. It has only been in the past three or four decades that the communications technology has existed to allow for the centralization and concentration that so characterize these enterprises. Decisions can be made in Zurich, New York or London, flashed around the world in minutes and implemented in Karachi, Manila, Lima, Nairobi, and a hundred other cities, towns and villages.

The speed and flexibility of their operations can have a dramatic effect on both world and national economies. Citing currency crises as one example, Tugendhat explains that companies, while not causing the crises, can exacerbate them through the ease with which they can move money in and out of a country. He points to the movement of $2,500 million out of the United Kingdom in a week, just before the pound was floated in 1972.[7] According to Brandt, many of the concerns about TNCs result from their ability to

> . . . race ahead in global operations out of reach of effective controls by nation-states or international organisations . . . to benefit from economic disorders at a time when many nations have suffered from them; and that they constitute a network of transnational power which has provided a new element in the struggle of political and economic forces.[8]

Armoured with the economic power, political power and even social power as a result of their ability to influence attitudes and purchasing decisions, is not that big a leap to draw the conclusion that TNCs could almost be considered the new sovereign states. Indeed, Max Gloor, a former director of the Swiss-based TNC, Nestlé, touched on this point:

> We cannot be considered either as pure Swiss, or as purely multinational, i.e. belonging to the world at large, if such a thing does exist at all. We are probably something in between, a breed on our own. In one word we have the particular Nestlé citizenship.[9]

The tensions and interactions between and among the two major actors on the international stage—governments and TNCs—are of a high degree of complexity. There is another factor, however, that is all too often overlooked in any examination of political and economic activity. The basic starting-point should be the needs of people, and the people themselves.

The Reagans, Thatchers, Gorbachevs, Mugabes or Gandhis of this world, while commanding a great deal of authority, can by no stretch of the imagination be regarded as commanding the full support of the people whose lives they attempt to direct. Similarly, despite advertising claims to the contrary, not everyone wants or needs a Coke, a Ford car, an IBM computer or a Philips video recorder.

The fundamental flaw in this two-system view of the world is its 'top-down' management. 'Solutions' to problems are imposed; people adjust to the solutions; or, if they do not, are characterized as deviants to the main stream of society. The existing models for the world are authoritarian, elitist and clumsily bureaucratic. Devising systems to enforce the rules has become more important than examining whether the rules or, for that matter, the 'game' itself are appropriate. Viewed in developmental terms, they are totally inappropriate, as the statistics mentioned earlier demonstrate.

Development is an unfolding process, an organic activity, and fundamentally relates to human beings, not to economic or political systems. Governments and economic units should be the servants of society, not the masters. It is, however, unrealistic to expect that the 'masters' will voluntarily give up authority and power, or that amongst themselves they will begin a negotiation process that will lead to a different structure of the world. For this to occur, a third system is required—the organization of the people themselves. In recent years, particularly in industrialized countries, but increasingly in developing countries, this third system has been growing. Its visible outcroppings are represented by countless pressure groups, consumer organizations, special interest groups, development agencies and other voluntary organizations.

Collectively, these are the non-governmental organizations (NGOs) which represent a diversity of viewpoints, ideas, needs and demands. Into an already complex world society they have added a third dimension and increased the complexity. However, in the same way that a three-dimensional model provides a better understanding than a two-dimensional drawing, the third dimension in international affairs added by the NGOs provides a more realistic framework for development.

NGOs lack the authoritative power of governments; they fall well below the economic clout of TNCs. But in terms of what Peter Willetts calls 'legitimacy', NGOs are the leading actors. Willetts explains that

> legitimacy attaches to actors, in the way we perceive their status, and to ideas or policies, in the way we measure their moral value . . . The ability to mobilise legitimacy should also be seen as a power capability. In this domain, it is conceivable that Amnesty International, for example, has greater power than any single government. It derives global legitimacy both from its very high status, one recognition of which was the award of the Nobel Peace Prize in 1977, and from the high moral value that so many people attach to the policies it is pursuing.[10]

Another striking example of the legitimacy and power of NGOs has been demonstrated in the decade-long campaign to improve infant feeding practices and to control the inappropriate promotion of artificial infant feeding products. Based on the direct experience of health workers in developing countries, first reported in scientific journals during the 1960s, the problems related to bottle feeding became a public issue in the early 1970s as NGOs took it up. The issue moved across nations, across continents, involved detailed research, copious publications, generated mass-media coverage including several television documentaries, led to two lawsuits, several government hearings, and captured the attention of the United Nations system. By 1981, an international code to control marketing practices had been devised and approved by the World Health Assembly (WHA), the governing body of the World Health Organization (WHO).

Beginning with a handful of organizations, there are now more than 100 groups in some sixty countries, linked together in a loose network, encouraging changes in public policy, in the behaviour of TNCs, and in health care practices that reflect the needs of people. The campaign, in addition to leading to some direct beneficial improvements in health, has led to a better understanding of the way in which TNCs operate and interact with governments, and how that interaction can be counterbalanced.

The organization of the campaign provides a new model for cooperation on important social, economic and political issues. The campaign has helped shatter many of the accepted myths about the role which TNCs play in the development process. It has led to a more fundamental questioning of organizational patterns and interactions between groups and individuals. It has shown that there are opportunities for people to participate more fully in making the decisions that affect their lives.

According to one commentator, the controversy over baby milks 'has all the ingredients of a classic political thriller'.[11] This book examines those ingredients and offers a fresh perspective on the controversy, not in terms of the scientific evidence behind the issue, but its organizational potential as the foundation for a new approach to global politics.

Notes

1. A. Toffler, *The Third Wave*, London, Pan Books, 1980, p. 93.
2. C. Tugendhat, *The Multinationals*, London, Penguin, 1973, p. 24.
3. E. Galeano, *The Open Veins of Latin America*, New York, Monthly Review Press, 1971.
4. W. Brandt *et. al.*, *North-South: A Programme for Survival*, London, Pan Books, 1980, p. 50.
5. United Nations, *Towards a World Economy that Works*, New York, UN Department of Public Information, 1980, p. 14.
6. Brandt, op. cit., pp. 187–8.
7. Tugendhat, op. cit., p. 13.
8. Brandt, op. cit., pp. 189–90.

9. Tugendhat, op. cit., p. 22.
10. P. Willetts (ed.), *Pressure Groups in the Global System*, London, Frances Pinter, 1982, pp. 23–4.
11. J. Bellini, 'Behind the Closed Doors of Nestlégate', *Vision*, May 1981, p. 10.

1. Identifying the problem

> It is clear to all but those who will not see that informed, adequate and relatively safe bottle feeding must *follow*, or at least accompany, but never *precede*, literacy, education, infection-free water supplies, sanitation and a standard of living which permits the purchase of enough baby foods, equipment and the means of sterilisation.
>
> Dr James Farquhar, consultant paediatrician at the Royal Infirmary of Edinburgh[1]

A vast upheaval in infant feeding habits has occurred in little more than 100 years. For nearly two million years infants have been successfully breast fed. However, by the middle of the twentieth century, more and more infants were being switched to the bottle, filled with specially-formulated, powdered skim milk preparations.

The shift occurred first in the industrialized countries of Europe and North America. In the United States in 1946, 38 per cent of infants were breast fed upon discharge from hospital. By 1966, the figure had dropped to only 18 per cent. In Sweden in 1944, just under 90 per cent of babies were breast fed at two months of age; by 1970 the figure had dropped to under 40 per cent. In Poland, over 80 per cent of three-month old infants were breast fed in 1937; by 1971, the figure was 30 per cent. In the United Kingdom, 40 per cent of three-month-olds were breast fed in 1947; by 1968, it was just over 10 per cent.[2]

This dramatic pattern soon spread to the developing countries of Asia, Africa and Latin America. In Singapore, over 80 per cent of babies were breast fed at birth in 1951; by 1971, the figure had dropped to just under 30 per cent. In Brazil, 96 per cent of babies were breast fed at one month of age in 1940; by 1974, only 39 per cent were breast fed.[3]

It was in the developing countries that the health consequences of artificial feeding were noticed most dramatically. There are several prerequisites to safe bottle feeding, most of them absent in developing countries;

— a clean water supply;
— fuel and facilities to boil the water and sterilize the feeding utensils;
— adequate income to be able to afford the milk powder;
— a level of literacy that allows for the mixing and sterilizing instructions to be carefully followed.

In Colombia, for example, in 1976 only 29 per cent of the rural population, and 72 per cent of the urban population had access to the potable water required for mixing with the powdered formulas.[4] In the capital of Malawi in 1972, 66 per cent of homes had no washing facilities, while 60 per cent had no indoor kitchen.[5] In Ethiopia during 1976, the cost of artificially feeding an infant for six months ranged between 50 and 140 per cent of the per capita gross national product (GNP), depending on the cost of formula used. In India, the range was between 40 and 60 per cent of per capita GNP.[6] In the Yemen Arab Republic in 1978, adult illiteracy rates for men were 74 per cent and 98 per cent for women.[7]

Bottle feeding in circumstances like these can lead to a lethal mix of weak contaminated fluids, which in turn, leave infants with gastroenteritis, diarrhoea, attacks of vomiting, dehydration, malnutrition and even death. Studies from around the world provide the grim statistics. Bottle-fed infants were three times more likely to die than breast-fed infants in Chile.[8] In Papua New Guinea, 23 per cent of artificially-fed infants were severely malnourished, compared to 3 per cent of breast-fed infants.[9] In Peru, the risk of malnutrition was nearly four and a half times higher for artificially-fed infants.[10] Moises Behar, chief of nutrition at WHO, concluded, following a study of infant health in Latin America, that

> where poor socio-economic and cultural conditions prevail, both early weaning and artificial feeding of infants are dangerous practices. They limit a child's natural defenses against enteric infections and greatly increase his(/her) exposure to an unhealthy environment.[11]

Poor socio-economic conditions exist in the industrialized nations as well. In the United States, a study conducted in a Navajo Indian reservation in the early 1960s found ten times more hospitalization of bottle-fed infants with gastro-enteritis and pneumonia than of breast-fed infants.[12] A similar study conducted among Indians in Canada in 1962 concluded that infant mortality rates could have been reduced by one-third if babies were breast fed. Another study in Canada in 1979 found that Indian babies who were fully bottle fed were hospitalized with infections ten times more often and spent ten times more days in hospital than fully breast-fed infants.[13] At Los Angeles County General Hospital, which serves a predominantly low-income Hispanic population, two to three bottle-fed infants per week were admitted for emergency treatment of severe diarrhoea during 1981.[14]

Nor is it only in poor socio-economic conditions that the choice of infant feeding makes a difference in health. In middle-class conditions in upper New York State during 1976, a comprehensive study found bottle-fed infants three times more susceptible to infections than breast-fed infants.[15] A study of well-nourished Japanese babies in Iwate showed that respiratory illness was nearly twice as common amongst bottle-fed infants.[16]

One researcher reviewed the available studies in industrialized countries from 1900 to the late 1970s and found similar results in England, France, Hungary and the United States.[17] A whole host of other complaints associated with bottle-fed infants are now clearly documented, including allergies to cow's milk, sudden infant death syndrome (cot death), obesity and dental caries.[18] This has led one American doctor to comment:

> The bottle-fed human baby is substantially more likely to suffer a whole nightmare of illnesses: diarrhoea, colic, gastrointestinal and respiratory infections, meningitis, asthma, hives, other allergies, pneumonia, eczema, obesity, hypertension, atherosclerosis, dermatitis, growth retardation, hypocalcemic tetany, neonatal hypothyroidism, necrotising enterocolitis, and sudden infant death syndrome. From a scientific, biological standpoint, formula feeding cannot be considered an acceptable alternative to breast feeding—especially since more than 99 per cent of new mothers are perfectly capable of doing it.[19]

As the volume of research on infant feeding steadily increases, so the evidence grows that inexorably points to the overwhelming benefits of breast feeding. Breast milk is a 'live' substance, containing enzymes, antibodies and immunizing agents specifically designed to protect the infant's very vulnerable digestive tract.[20] In addition to this protection, it provides the ideal form of nutrition. One researcher has noted that the change in the composition of the milk from feed to feed and even during the course of a single feed may help develop the baby's appetite control, thus safeguarding against obesity.[21] On top of all this, breast milk requires no mixing or sterilization, is always at the right temperature, and is virtually free of charge.

Breast feeding has been linked to an improved emotional and psychological bond between mother and child, leading to a more secure and supportive start in life for the baby and, in all probability, a more contented and responsive child.[22] For the mother, breast feeding has been found to provide satisfaction and pleasure—a natural continuation of the entire reproductive process—intercourse, childbirth and breast feeding.[23] Breast feeding also helps the mother return naturally to her pre-pregnant condition through stimulating the gentle contraction of her womb and using up the layer of surplus fat that is generally laid down during pregnancy.[24]

Breast feeding can also help birth spacing, providing it is practised on demand and exclusively for the first few months of a baby's life. Evidence from several studies shows that breast feeding can prolong the onset of ovulation for anything up to two years.[25] It should be noted that, on an individual level, breast feeding is not necessarily a reliable contraceptive, but on a society-wide basis it has been estimated to provide more 'women-years' of protection against pregnancy than the total contraceptive protection achieved through other family-planning programmes.[26] Particularly in developing countries, this natural birth spacing can have a profound impact on the nutritional status of the mother and her children.

Reasons for the decline in breast feeding

With so much information now available on the benefits of breast feeding and disadvantages of bottle feeding, the widespread shift of feeding patterns seems inconceivable. However, much of the information has only recently become widely known. For most of the twentieth century, the emphasis throughout society has been to develop more 'scientific and technological' solutions to problems, and it was this thrust which influenced attitudes about infant feeding.

Health care professionals have played a key role in this, largely by omission rather than commission. Their advice to mothers has often been minimal and confused, through a lack of knowledge or interest in nutrition generally, including breast feeding. One study of 742 nursing schools throughout the world found that nutrition information was almost always inadequate and that cultural attitudes, the composition of locally available foods, and particularly the significance of breast milk, were rarely covered in the curriculum.[27]

In 1976, a detailed study in Manila found that only 33 per cent of doctors and 6 per cent of midwives understood the basic reflexes controlling lactation. The researcher commented that 'although most respondents said they were in favour of breast feeding, in fact, only a few had the knowledge and attitudes needed to actively encourage the practice'.[28]

The increasing mechanization of childbirth, with the management of deliveries and fixed routines in hospitals, has compounded the problem. A survey in the United Kingdom found that mothers who had normal deliveries—free from anaesthesia and induced labour—and were given their babies to hold and feed within four hours of delivery were more likely to begin breast feeding and more likely to continue. It also noted that routine bottle feeding was the norm, with 86 per cent of all breast-fed babies also receiving artificial feeds during the first week of life.[29] The survey noted that

> breast feeding is difficult to establish if artificial feeds are given routinely . . . Mothers also reported that bottle feeding, when they were trying to breast feed, undermined their confidence in their capacity to produce milk—an important point . . . This influenced some mothers to stop breast feeding.

Confidence is one of the most important factors in successful breast feeding. While the milk itself is produced largely as a result of hormone-triggered chemical interactions which are virtually automatic, the delivery of that milk is dependent on another hormone which stimulates the 'let-down reflex'. This reflex can be affected by psychological or emotional factors—stress and anxiety.[30] When this occurs, little milk flows to the child; the baby is left hungry; the mother's breasts become sore and engorged, and the possibility of cracked nipples and breast abscesses increases.

Thus, while a tiny proportion of mothers are physically unable to produce enough milk—one estimate puts it at under 1 per cent[31]—many more mothers claim insufficient milk as the main reason for switching to the bottle. The British survey found 61 per cent of mothers who stopped breast feeding by the baby's sixth week citing insufficient milk.[32] The same reason was given by 43 per cent of mothers in Kingston, Jamaica[33] and 31 per cent in South Brazil.[34]

All three studies expressed doubts about the validity of these findings. The

British survey said, 'it seems unlikely that such a high proportion of mothers as was found in the survey were really incapable of producing enough milk for their babies'. The Brazil study, which questioned mothers further, found that two-thirds went on to say that the milk was 'weak, thin, translucent'—precisely the way it should look! In Jamaica, 'many mothers who said they had insufficient milk obviously had enough when questioned further'.

The lack of confidence itself stems from many factors in society. There is very little formal public education which sets out the pros and cons of infant feeding. The social support system, traditionally the extended family, is disappearing in many communities, leaving a young mother with no source of advice. Similarly, the trend towards increased urbanization, with more people moving to cities and facing a faster pace of life with more pressure and strain, has further isolated mothers. The urban life-style has also created changes in work patterns, with more women taking up paid employment in circumstances where breast feeding is difficult. The increased perception of breasts as sexual objects has led to feelings of embarrassment for many women attempting to breast feed in public. This has been reinforced by the concept that the feeding bottle is the modern, convenient and liberating answer. While none of these factors has a particularly significant effect when viewed in isolation, their compound interaction has produced the profound changes in infant feeding habits.

Underlying and reinforcing many of these factors has been the commercial promotion of artificial infant feeding and the requisite products. Originally designed as a life-saving tool for use when breast milk was not available, in just over 100 years infant formula has become a high-turnover high-profit consumer product.

One commentator points out that the formula manufacturers have played an active role in creating demand for their products, thus invalidating the belief that any free market governed by demand and supply exists in this case. He summarizes the situation in this way:

> By and large, it can be said that the infant-milk companies have intervened in the natural relationship between a mother and her child in such a way as to modify it by introducing their product. As a result, the consumption level of the formula products (or use of bottle feeding) is strongly influenced by companies' marketing activities.[35]

The range of activities has been diverse:

- direct advertising through radio, television, press, mass-circulation magazines, outdoor billboards and posters, and a host of give-away items (T-shirts, nappies, bibs, aprons, feeding bottles, carrier bags) which promote the products;
- providing mothers with booklets, leaflets and brochures which advertise the products as well as giving advice on infant care;

— providing health workers with promotional materials, gifts, lunches, trips and samples for their own infants;
— making samples available for mothers, either directly by having sales representatives, often former nurses, give the samples, or indirectly by leaving them at hospitals and clinics for the medical staff to pass on.[36]

The messages which come across in this promotion fuel the concerns, the fears and the hopes prevalent in society, and, by strengthening the doubts that exist about breast feeding, serve effectively to undermine it.

The possibility of insufficient milk is reinforced with the headlines in advertising for Nestlé products: 'When mother's milk fails . . .'.[37] The new, modern and scientific approach to infant feeding is reinforced by Glaxo's 'Give your baby the benefit of modern research—Vitamilk'.[38] Providing the best for baby is guaranteed by Cow & Gate: 'You can't do better for your baby';[39] or by Abbott/Ross who advise that 'all your baby needs for healthy growth is love and Similac'.[40] Convenience and the changing role of women in society are highlighted in another Nestlé headline: 'Now Dad can give baby all the goodness of mother's milk'.[41]

A communications specialist at Stanford University examined the most recent publications of all the major infant-formula manufacturers in the United States during 1980 and found five themes running through the literature:[42]

— formula is supported by medical experts;
— it is supported by science;
— it makes for healthy babies;
— it is fed by loving (and lovely) mothers;
— it fights mother's fears.

Taking over the dissemination of information to mothers is only half the picture. A medical seal of approval for the products is equally important and, here again, the manufacturers have spared no cost to inundate health workers with information and gifts.

One company, Abbott/Ross, offers a complete 'Similac Hospital Formula System', which by 1968 was in use in more than 2,500 maternity hospitals in the United States. According to the company, it was reaching over half the newborn infants.[43] The system provides nine different sample packs for mothers: a tape measure, products catalogue, teaching aids and information materials, newsletters, name tags and training for hospital staff in the use of the system and the formulas; and for the hospitals themselves Abbott/Ross maintains a hospital planning department staffed by experienced planning experts. According to one commentator, Abbott helps design over 200 maternity departments a year in the United States.[44] The layout of these departments makes breast feeding difficult. Mothers are separated from their babies and the nursing staff can quickly and conveniently administer donated formula in ready-to-feed bottles.

Another form of 'assistance'—a 'kick-back' scheme—was uncovered by the

American television network CBS in the Dominican Republic in 1978. Abbott, Bristol-Myers, Nestlé and the local subsidiary of Wyeth all formed a pharmaceutical association which agreed to pay the Dominican Republic Medical Association one-half of 1 per cent of sales of all pharmaceutical products, including infant formula. It was a type of 'sales commission' that in 1977 totalled $80,000.[45] During 1981, Nestlé provided nearly $250,000 for speakers and food at seminars and conventions in the Philippines, including $1,250 for the annual convention of the Philippine Paediatric Society. The majority of the twenty-two seminars sponsored by Nestlé were for midwives.[46]

The companies know that persuading the medical profession works. Cow & Gate's annual report pointed to its highest-ever exports in 1976, thanks to 'wide contacts with the medical profession'.[47] As long ago as 1970, *American Drug* magazine noted the effectiveness of this approach: 'Baby formula is a high-volume item that is practically pre-sold to new mothers via physician/hospital endorsement'.[48]

Working for change

Identifying the complex causes of a problem is the first step towards its resolution. The second is identifying possible solutions, and those who have the power to encourage and implement them. In the case of infant feeding, the solutions, like the problem, run through all levels of society, across national boundaries and involve a varied group of actors.

Simply announcing the now widely accepted premise that breast milk is best for infants will not change a great deal. The Nutrition Foundation of the Philippines came to that conclusion:

> In connection with the breast-feeding message, we could do well to look into how we can help mothers become able to breast feed instead of merely harping on the virtues of breast feeding. If legislations are required (such as for the control of advertising and promotion 'gimmicks' of milk companies, the establishment of rooming-in in all hospitals, etc.), then those in the campaign must be willing to work for such. Only when we become creative in our approaches can we make headway in the 'Back to Breast' campaign.[49]

Elisabet Helsing, a nutritionist with the Norwegian Directorate of Public Health and a founder member of the Norwegian breast-feeding mothers' organization, Ammehjelpen, confirms the need for a multi-faceted approach. She examines the cultural/psychological, societal and material, economic and legal factors that are involved (see Table 1), and concludes that, without taking into account their effect, 'motivational' campaigns to encourage mothers to breast feed are likely to fail. She calls for a coherent policy:

Table 1 Factors influencing mothers' choice of mode of infant feeding

Cultural/psychological dimensions	Societal dimensions	Material, economic, and legal dimensions
Knowledge about infants' nutritional needs about the management of breast feeding Educational level Traditions on infant feeding on maternal feeding on child care Attitudes and perceptions of the mother, father, family, and society concerning: maternal role child rearing breast feeding paternal role sexuality female role (production responsibilities) society's responsibilities and obligations towards its members	Delivery practices and routines (including pre- and post-natal situation) traditional modern Social support systems for newly-delivered mothers traditional modern private public Pattern of household work (reproductive work) rapidly changing family structures male responsibility for social reproductive work Patterns of child care use and availability of child caretaker age and skill of child caretaker Control over resources material resources time Marketing practices of baby food industries through health care system through personal contacts through mass media	Family's income in cash and kind Family's social class Number and age of dependents Availability of child care price quality proximity to work-place Mothers' work situation at home availability of free time workload social obligations Mothers' employment situation job security flexibility of working hours distance from home to work-place nature of work Family's employment situation Maternity protection legislation theory practice Mothers' physiological status nutritional status age parity medical history Existence of an alternative to breast feeding availability feasibility price

Source: E. Helsing, 'Infant feeding practices in Northern Europe', *Assignment Children*, No. 55/56, UNICEF, 1981, p. 87.

> It must be materially possible for mothers to breast feed; societal factors such as maternity ward routines and social support systems must be conductive to breast feeding; and the promotion of alternatives must be regulated. Cultural/psychological factors to be considered concern above all the knowledge of health workers about the management of breast feeding.[50]

Ironically, most of the actors play a dual role in this controversy: they are both part of the problem and part of the solution.

Health workers are involved through their non-existent or negative advice on infant feeding, yet they have an important role in providing appropriate advice, support and in helping to remodel practices within health care facilities. Governments in developing and industrialized countries which fail to implement and enforce maternity legislation, advertising controls, or public and professional educational programmes are adding to the problem. Yet their authority and power are a necessary part of the solution.

People—mothers, fathers, grandparents, childen and all the others who form the nebulous 'general public'—who passively accept the situation contribute to the inertia that makes change difficult. When their voices are heard, either directly by standing up for their rights, or indirectly through the countless groups and organizations that advocate on their behalf, they add weight to the momentum of change. Similarly, the media that convey so many of the strong messages about infant feeding through 'editorial' coverage and paid advertising should be sensitized and involved in working for solutions.

The TNCs and national companies producing infant milks and feeding utensils exert a powerful influence that has been largely detrimental to infant health. Their activities and practices are, however, an integral part to any solution. Because it is an international problem, national solutions will not suffice. Ideas, attitudes, cultural practices know no boundaries. The Democratic Republic of Yemen took steps to improve infant feeding practices some years ago by stopping advertising for infant formulas. However, as one of the country's health officials notes, 'It is not possible to prevent neighbouring radio and TV stations from broadcasting such advertisements'.[51] Thus, international agencies and intergovernmental organizations need to put the infant feeding question on their agendas if they are to be part of the solution.

For all the complexity of the problems and solutions, the basic tools that are required to arrive at solutions are relatively simple and highly accessible. Information is a key tool—the basis of power. As information about the infant feeding issue spreads and an understanding of the problems and possible solutions grows, so the resistance to change becomes less sustainable.

Pressure is another important tool. It is somewhat intangible, but none the less real. Coupled with information, it provides the crucible from which political will and hence action can be formed. Pressure works best, however, when it is concentrated or directed at a specific point. If the left hand is bleeding, applying a tourniquet to the right leg is unlikely to be successful. This then leads to the third step in working for change—identifying the major target.

In the case of infant feeding, all the participants mentioned above are targets—the subject of pressure—as well as potential sources of pressure. However, through the decade of campaigning around this issue, one target has emerged as a clear focus. The group which had the most invested in, and the most to gain from, artificial infant feeding were the manufacturers—the national and transnational companies. This was the group that became the primary target.

Notes

1. A. Chetley, *The Baby Killer Scandal*, London, War on Want, 1979, p. 55.
2. D. B. Jelliffe and E. F. P. Jelliffe, *Human Milk in the Modern World*, Oxford, Oxford University Press, 1978, pp. 189–91.
3. Ibid., pp. 217–18.
4. US Senate, *Marketing and Promotion of Infant Formula in the Developing Countries: Report of the US Senate Health and Scientific Research Sub-committee*, Washington, DC, 1978, p. 118.
5. Chetley, op. cit., p. 50.
6. World Health Organization, *Contemporary Patterns of Breastfeeding: Report on the WHO Collaborative Study on Breastfeeding*, Geneva, 1981, p. 139.
7. J. Firebrace, *Infant Feeding in the Yemen Arab Republic*, London, Catholic Institute for International Relations and War on Want, 1981, p. 13.
8. S. J. Plank and M. L. Milanesi, 'Infant feeding and infant mortality in rural Chile', *Bulletin of the World Health Organization*, 48, 1973, p. 203.
9. J. Lambert and J. Basford, 'Port Moresby infant feeding survey', *Papua New Guinea Medical Journal*, **20**, 4, December 1977, p. 175.
10. P. A. Borgoltz and M. C. Latham, 'Effects of artificial feeding on nutritional status of the infant population of Lima, Peru', *American Journal of Clinical Nutrition*, **32**, 4, April 1979, p. 948.
11. Jelliffe and Jelliffe, op. cit., p. 287.
12. L. Salisbury and A. G. Blackwell, *An Administrative Petition to the United States Food and Drug Administration and Department of Health and Human Services*, San Francisco, Public Advocates Inc., 1981, p. 27.
13. Chetley, op. cit., p. 27.
14. Salisbury and Blackwell, op. cit., p. 38.
15. A. S. Cunningham, 'Morbidity in breast-fed and artificially-fed infants', *Journal of Pediatrics*, **5**, 726, May 1977.
16. Jelliffe and Jelliffe, op. cit., p. 261.
17. A. S. Cunningham, 'Breastfeeding and morbidity in industrialised communities: an update', in D. B. Jelliffe and E. F. P. Jelliffe (eds), *Advances in International Maternal-Child Health*, Vol I, Oxford, Oxford University Press, 1981, pp. 128–68.
18. Jelliffe and Jelliffe, *Human Milk in the Modern World*, 1978, op. cit., pp. 241–69.
19. R. S. Mendelsohn, *Confessions of a Medical Heretic*, New York, Warner Books, 1980, p. 156.
20. G. J. Ebrahim, *Breast Feeding: The Biological Option*, London, Macmillan, 1978, p. 55.
21. B. Hall, 'Changing composition of human milk and early development of an appetite control', *The Lancet*, 5 April 1979.
22. Jelliffe and Jelliffe (1978), op. cit., p. 267.

23. N. Newton, 'The importance of prenatal education', *Child and Family*, winter issue, 1969, p. 61.
24. Chetley, op. cit., pp. 33–4.
25. See, for instance, J. Knodel, 'Breast feeding and population growth', *Science*, **198**, 16 December 1977, p. 1111; *Sexual Medicine Today*, 'The neglected contraceptive', May 1978; R. Short, in 'Nature's own contraceptive', BBC Radio broadcast, 6 May 1980 (Tape No. TEH05/650S036).
26. Knodel, op. cit.
27. Jelliffe and Jelliffe (1978), op. cit., p. 343.
28. Chetley, op. cit., p. 81.
29. Department of Health and Social Security (UK), *Breast Feeding*, London, 1978, pp. 17–19.
30. Jelliffe and Jelliffe, (1978), op. cit., p. 22.
31. World Health Organization, *Provisional Summary Record of the Eighth Meeting of Committee A, Thirty-third World Health Assembly* (Doc. No. A33/A/SR/8), Geneva, 17 May 1980, p. 11.
32. Department of Health and Social Security, op. cit., p. 14.
33. S. M. Grantham-McGregor and E. H. Back, 'Breast feeding in Kingston, Jamaica', *Archives of Diseases in Childhood*, **45**, 404, 1970.
34. P. L. R. Sousa *et al.*, 'Patterns of weaning in South Brazil', *Journal of Tropical Paediatrics and Environmental Child Health*, August 1975, p. 210.
35. P. A. Borgoltz, 'Economic and business aspects of infant formula promotion: implications for health professionals', in D. B. Jelliffe and E. F. P. Jelliffe (eds), *Advances in International Maternal-Child Health*, Vol II, Oxford, Oxford University Press, 1982, p. 179.
36. See, for instance, IBFAN, *Breaking the Rules*, Geneva, Minneapolis and Penang, 1982; IBFAN, *Infant Formula Promotion*, Geneva, London and Minneapolis, 1981.
37. Poster for Lactogen formula, from a clinic in Zimbabwe in 1980.
38. Advertisement in the *Pakistan Times*, February 1981.
39. Advertisement in a babycare booklet in Malaysia, 1976.
40. Front cover of babycare booklet in Kenya, 1981.
41. Advertisement for Nan formula in a South African magazine, 1977.
42. Salisbury and Blackwell, op. cit., p. 94.
43. Borgoltz, op. cit., p. 186.
44. E. Baer, 'Babies means business', *New Internationalist*, No. 110, April 1982, p. 23.
45. *CBS Reports*, transcript for documentary 'Into the mouths of babes', broadcast 5 July 1978, New York, pp. 17–20.
46. IBFAN (1982), op. cit., p. 31.
47. Borgoltz, op. cit., p. 177.
48. Ibid., p. 186.
49. Nutrition Foundation of the Philippines, Editorial 'On the breast feeding campaign', *Bulletin of the Nutrition Foundation of the Philippines*, **20**, 2, March/April 1980, p. 2.
50. E. Helsing, 'Infant feeding practices in Northern Europe', *Assignment Children*, No. 55/56, Geneva, UNICEF, 1981, p. 88.
51. World Health Organization, *Thirty-Fourth World Health Assembly, Summary Records of Committees* (Doc. No. WHA34/1981/REC/3), Geneva, 1981, p. 186.

2. Successful selling

> Despite the somewhat unfavourable economic situation in Latin America, intensive promotional and advertising efforts brought about a marked improvement in sales of our brands of milk products.
>
> Nestlé, *Annual Report*, 1963.

The infant milk market is dominated by two types of TNCs—agribusiness or food manufacturers and pharmaceutical companies. More than thirty companies are involved; about one-third are pharmaceutical producers, the rest concentrate on food products. However, some twelve companies control the bulk of the world market, estimated at US$3.3 billion in 1983. Of these, the market leader is Swiss-based Nestlé with nearly one-third of world sales, while the top four companies control two-thirds of the market (see Table 2).

This intense concentration is reflected in national markets, with generally 90 per cent or more of the market controlled by three or four companies. In twenty-four selected countries (see Table 3), the dominance of Nestlé, Abbott/Ross, Bristol–Myers and Wyeth is clearly shown.

Three characteristics develop from the oligopolistic nature of the infant milk market: a lack of price competition, emphasis on product differentiation, and high promotional expenditure to maintain market share.[1] With this type of market, there is little benefit to the consumer. Companies rarely start a price war, recognizing that any increase in sales from lower prices will be offset by a decreased profit margin. Product differentiation is largely fictional. The British Department of Health and Social Security concluded in 1980 that there was 'no firm evidence upon which to base a choice of one particular infant formula as being more suitable than another'.[2] Thus the 'real' difference between the competing products depends on the marketing skills of the different companies.

To understand how this mechanism works it is necessary to examine the changing character of advertising and marketing—both now largely controlled by TNCs. Of the top fifteen advertising agencies, fourteen are American-based, with the leading agency based in Japan. The eight largest American agencies operate in more than twenty countries each, with one agency, McCann Ericksson, operating in more than fifty countries.[3]

The growth of these giant concerns has followed the growth of TNCs generally, with agencies often expanding into a new country after one or two of their major clients have entered a new market. The basic imperative they followed was explained by *Business International* in 1963: 'The most vital challenge facing today's international firm is no longer how to build up the production of goods, but how

Table 2 Leading transnational companies in the infant milk sector (ranked by infant milk sales, 1979, with estimated 1983 sales)

Rank company	Country of origin	Total sales 1979 (US$m.)	Infant milk sales 1979	Estimated 1983 infant milk sales*	Percentage of market (cumulative)
1. Nestlé	Switzerland	14,113	700	1,025	31.1(31.1)
2. Abbott/Ross	USA	1,683	335	490	14.9(46)
3. Bristol–Myers	USA	2,753	250	366	11.1(57.1)
4. American Home Products/Wyeth	USA	3,649	240	351	10.6(67.7)
5. VARTA	FRG	900†	125†	201	6.1(73.8)
6. Glaxo	UK	1,080	90†	145	4.4(78.2)
7. Morinaga	Japan	1,227	65†	105	3.2(81.4)
8. Nutricia	NL	230‡	65†	105	3.2(84.6)
9. Meiji	Japan	1,438	50†	80	2.4(87)
10. Cow & Gate	NL	n.a.	50†	80	2.4(89.4)
11. BSN Gervais Danone	France	3,866	40†	65	2 (91.4)
12. CDC	Canada	2,015	25†	41	1.2(92.6)
13 Snow Brand	Japan	3,239	15†	24	0.7(93.3)
14. Syntex	USA	471	15†	24	0.7(94)
Others				198	6 (100)

Notes:
* Estimated sales for 1983 based on a 10 per cent per annum growth rate in sales, the lowest figure suggested by marketing expert James Post at the US Senate Hearings, May 1978.
† 1978 figures.
‡ 1977 figures.

Source: Adapted from P. A. Borgoltz, Ch. 12, in D. B. Jelliffe and E. F. P. Jelliffe, (eds), *Advances in International Maternal-Child Health*, Vol. II, Oxford, Oxford University Press, 1982, p. 160.

to increase their penetration.'[4] One of the prime tools TNCs used to do this was advertising.

This works in two ways: first, by displacing local production, if there is any, through high-profile media presence; and, second, through introducing new consumption patterns for goods.[5] In developing countries particularly, this can have a detrimental effect. Janus and Roncagliolo, who studied the impact of advertising in Latin America, concluded that the use of national wealth to advertise less than essential consumer goods, when basic human needs for food, clothing and shelter were not being met, was 'highly questionable'.[6] They also point with concern to the undermining of 'cultural sovereignty' through the dissemination on a global scale of a specific set of cultural patterns, a particular 'way of life'.

However, two commentators note that 'advertising does more than merely sell products and form consumption patterns: it informs, educates, changes attitudes and builds images'.[7] Selling ideas, not products, is the function of advertising

Table 3 Concentration of infant formula markets in 24 selected countries, with estimated 1983 sales

Country	Companies share of market (%)				Market value of sales		Estimated sales in 1983*
	Top 1	Top 2	Top 3	Top 4	Year	US$m.	
Brazil	95(N)				1978	70	113
Colombia	70(N)	97(W)	100(A)		1976	20	39
Ethiopia	40(N)	65(Mo)	85(Me)	95(A)	1975	1.5	3.2
FRG	50(M)	85(N)	95(H)		1977	125	221
France	65(N)	95(BSN)			1978	110	177
Hong Kong	n.a.	n.a.			1978	17	27
India	35(Am)	51(G)	65(N)		1975	51	109
Indonesia	60(SH)	80(N)			1982	50	55
Malaysia	40(N)	80(CCF)	90(D)		1977	20	35
Mexico	55(N)	80(local)			1978	50	81
Nigeria	40(N)	59(W)	77(CG)	90(A)	1975	50	107
Panama†	75(N)	87(A)	95(BM)		1978	20	32
Peru	65(N)	90(W)	98(A)		1976	10	19.5
Philippines	47(N)	85(W)	99(BM)		1975	22	47
Singapore	48(N)	68(W)			1981	21	25
South Africa	72(N)	90(W)			1979	65	95
Sudan†	n.a.	n.a.			1980	3	4
Sri Lanka	32(N)	58(local)	83(G)		1978	20	32
Tanzania	80(N)	89(W)	98(A)		1977	2.5	4.4
Thailand	66(N)	75(W)	83(BM)	88(D)	1979	36.5	53
UK	42(W)	75(CG)	98(G)		1980	72	96
USA	55(A)	90(BM)	99(W)		1981	700	847
Yemen Arab Rep.	60(N)	80(Mo)			1980	18	24
Zimbabwe	95(N)	100(W)			1980	3	4

Key to Companies: N = Nestlé, W = Wyeth, A = Abbott/Ross, Mo = Morinaga, Me = Meiji, M = Milupa, H = Humanische Milch, BSN = BSN Gervais Danone, Am = Amul, G = Glaxo, CCF = Friesland, D = Dumex, CG = Cow & Gate, BM = Bristol-Myers, SH = Sari Husada

Note:
* Estimated sales for 1983 based on a 10 per cent per annum growth rate in sales, the lowest figure suggested by marketing expert James Post at the US Senate Hearings, May 1978.
† Based on import figures only.

Source: Adapted from P. A. Borgoltz, Ch. 12, in D. B. Jelliffe and E. F. P. Jelliffe (eds), *Advances in International Maternal-Child Health*, Vol. II, Oxford, Oxford University Press, 1982, p. 163; and supplemented with additional information from various sources.

according to one marketing manager, who sees advertising agencies' main role as attempting 'to influence human behaviour in ways favourable to the interests of their clients'.[8] The influence sought over economic behaviour is fairly obvious: the development of a consumption ethic. Advertising urges the individual to buy, buy, buy; all in the interest of performing a 'patriotic service' by keeping the economy going.[9]

In developing countries, such an imperative can have a profound effect on basic strategies for development. By stimulating demand for non-essential goods, it pressurizes governments into giving priority to the investments necessary to produce them, to the detriment of the basic needs of the population. Advertising thus constitutes a parallel and non-official development programme.[10] An illustration of this is provided in Table 4 which contrasts some of the objectives agreed for the International Development Strategy for the Second Development Decade approved by the UN in 1970, with some of the fundamental premises of advertising.

Table 4 Comparison between two development models

United Nations International Development Strategy: some objectives	Transnational consumer model
1. An attempt to improve the quality of education, reduce illiteracy and reorient programmes must be made at all levels, with a view to satisfying the needs of development.	1. Advertising is an alternative form of education in so far as it promotes social values and develops methods which permit the incorporation of illiterate and semi-literate populations into the consumer society.
2. All developing countries must formulate coherent health programmes for the prevention and treatment of illness and raise the general level of the population's health and well-being.	2. It has been proven that advertising, as an instrument of transnational pharmaceutical corporations, frequently promotes medical products which reflect primarily market interests rather than the genuine health of the population.
3. The level of nutrition must be improved, in so far as concerns the average intake of calories as well as the protein content.	3. There is evidence of a strong link between transnational advertising and the promotion of food and beverages lacking in nutritional value, which in certain cases may even be damaging to the population's health.
4. Strong efforts must be made to promote private savings through financial institutions, post office savings and other systems.	4. The fundamental task of advertising is to encourage the spending of all available disposable income on consumer goods, with a view to increasing and accelerating the sales of the firms' products.
5. Developing countries will adopt national policies which promote the participation of children and teenagers in the development process and guarantee that their needs are met in an integral manner.	5. Children and teenagers are a lucrative market and, consequently, they represent an important target group to which consumer models, superfluous and culturally alien to their national frameworks of development, are directed.

Source: N. Janus and R. Roncagliolo, 'Advertising, mass media and dependency', *Development Dialogue*, 1, 1979, p. 95.

Management theorist Peter Drucker takes a different view, as in fact do many other business commentators and the TNCs themselves. Drucker claims that good marketing skills make possible 'economic integration and the fullest utilisation of whatever assets and productive capacities an economy already possesses'.[11] He says this contributes to the greatest needs, 'the rapid development of entrepreneurs and managers'. He further defines marketing as 'the process through which economy is integrated into society to serve human needs'.[12]

TNCs are seen as the main vehicles for this. A vice-president of the Monsanto company claims that 'multinationals perform a unique role in our society. They came into being via the genius of self-interest serving needs by bringing together resources in a unique way that no other societal organisation has matched'.[13] While the basis for that statement is true, what goes unquestioned is the assumption that this 'unique role' is the *best* role.

Nestlé puts the argument in even stronger terms in a glossy 228-page report describing its activities in developing countries. 'While Nestlé is not a philanthropic society, facts and figures clearly prove that the nature of its activities in developing countries is self-evident as a factor that contributes to economic development.'[14] Susan George carefully studied the report and found no 'facts and figures' that could either confirm or deny the assertion.[15] What she did find was that Nestlé's 'support for development' benefited the already established elites in Latin America; that prices paid for local raw materials were often below world market rates; and that, no matter what the basic needs of the country might be, Nestlé introduced the manufacture of its standard product range. She concluded that Nestlé's contribution to development was 'at best, a matter of enlightened self-interest'.

What the publicists for TNCs often overlook is the fact that economic development is not automatically beneficial to the mass of the population. One comprehensive analysis of forty-eight different indices of social, political and economic characteristics in seventy-four countries found that 'hundreds of millions of desperately poor people have been hurt rather than helped by economic development . . . There is no automatic, or even likely trickling down of the benefits of economic growth to the poorest segments of the population in low-income countries'.[16]

Nor is much consideration given to the fact that TNCs, for all the investment and enterprise they undertake, concentrate those resources in very few countries. The Brandt Report notes that 70 per cent of TNC investment has been in only fifteen developing countries—'mainly those which could offer political stability and a convenient economic environment'.[17] Another 25 per cent has gone into the oil-exporting developing countries.

TNCs and politics

The second element to TNC activity which is buttressed by marketing and advertising is the influence sought over political activity. As part of the TNC community, advertising agencies promote not only the ideas of consumerism and the products to be consumed, but serve to promote and justify the positive role of TNCs in development and, indeed, the entire free enterprise system.

Alfred Hollender, chairman of the International Advertising Agency told the 24th World Congress of Advertisers in 1972,

> The multinational corporation is a constructive international force, contributing significantly to the economic development and improved standards of living in the countries where it operates. The multinational corporation is the most effective instrument for this purpose that society has yet devised. And yet the danger signals are everywhere, with the role of multinational corporations questioned in many countries and in the United Nations.[18]

Less than three months later, Salvador Allende, president of Chile, provided the General Assembly of the United Nations with good reason to question the role of TNCs.

> Two firms that are part of the central nucleus of the large transnational companies that sunk their claws into my country, the International Telegraph and Telephone Company (ITT) and the Kennecott Copper Corporation, tried to run our political life . . .
>
> Last July, the world learned with amazement of different aspects of a new plan of action that ITT has presented to the US Government in order to overthrow my Government in a period of six months . . . They wanted to strangle us economically, carry out diplomatic sabotage, create panic among the population and cause social disorder so that when the Government lost control, the armed forces would be driven to eliminate the democratic regime and impose a dictatorship . . .
>
> Distinguished representatives, before the conscience of the World, I accuse ITT of trying to provoke a civil war in my country—the supreme state of disintegration for a country. This is what we call imperialist intervention.[19]

ITT's response was to launch a massive $6.4 million advertising campaign in 1974 to improve its image. By 1975, as a result of this, the number of people in the United States who believed that ITT 'cares about the general public' increased from 20 per cent to 43 per cent.[20]

Not surprisingly, TNC leaders attempt to play down the involvement of ITT in Chile's internal politics and to claim that TNCs have no political influence. Eric Gabus, a Nestlé manager, says:

> In a well-organised democratic society, as are all the industrialised societies where MNCs are formed and exercise the bulk of their activities, the MNC has

> no direct political power. At most it is a partner in the political dialogue, although it is sometimes privileged by its economic responsibility.[21]

He calls the intervention of ITT in Chile an 'exceptional incident'. Exceptional it may well be in terms of the extent of the intervention. But TNC interference with political decisions in national settings is by no means a rare occurrence, nor is their political power inconsequential.

When Sri Lanka attempted to rationalize its national drugs policy in the early 1970s, the country faced strong opposition from TNCs, particularly those that were American-based. One company, Pfizer, stalled negotiations over local production plans for four years until the Sri Lankan government, in frustration, suggested that it might nationalize the company. The Sri Lankan prime minister was then visited by the United States' ambassador who indicated that the supply of American food aid to Sri Lanka would be put in jeopardy by such action. Sri Lanka backed down.[22]

In Brazil, the West German embassy delivered a memo to the Brazilian government protesting about inadequate price increases for drugs. It noted that German drug companies would have some difficulties with this and warned that Germany's cooperation in other areas of Brazil's industrialization plans could be affected.[23]

Anthony Sampson documented the economic and political power of the TNCs in the oil industry:

> As the business became more global in the forties and fifties, so governments thought they were using their companies by encouraging them abroad, with antitrust clearances, tax advantages and diplomatic support, while the companies were in fact far better at using them, in ways that were often against their governments' interests.[24]

These, and many other examples that dot the world, demonstrate a basic underlying imperative of TNCs to encourage the development of political systems that are supportive of the free enterprise system. Drucker explains,

> without marketing as the hinge on which to turn, economic development will almost have to take the totalitarian form . . . If we want economic development in freedom and responsibility, we have to build it on the development of marketing.[25]

Professor John Ewing is more blunt: 'While it is perhaps not the primary responsibility of business to fight communism, this is certainly a secondary objective—or should be, since to be on the losing end is beyond acceptance'.[26]

Several clear perspectives on the thinking and motivation behind transnational enterprises emerge. Sales, profits and the maximization of capital are the unquestioned primary objectives. There is no dispute between TNCs and their critics on that point. The conflict occurs over how well matched those objectives are to basic social and economic development. TNCs push the line that they are obviously matched, and that the interests of people are adequately and appropriately catered to by the interests of TNCs. Any critical assessment almost inevitably shows that this is not always the case.

The TNCs claim either no political power or adopt a neutral political stance. Yet in both theory and practice the evidence shows this is not the case. As J. K. Galbraith points out, 'perhaps the oldest and wisest strategy for the exercise of power is to deny that it is possessed'.[27] He goes on to note that persuasion is the basic instrument for the exercise of power. It requires an image of economic life that is congenial to the organizations exercising power. Of vital importance, Galbraith says, is the instruction that implants this image. 'It persuades people that the goals of the organisation are really their own, or paves the way for such persuasion'.

Ivan Illich expresses concern about this persuasion not simply in terms of the way it shapes preferences, but in the way it influences people's sense of what is possible.[28] Although this is done under the guise of providing more choice, Illich claims that more and more people have fewer basic choices. The choices boil down to a restricted set of 'packaged goods and services' served up by business, the media, the health care systems, governments and educational systems.

Nowhere is the packaging of a particular set of goods and services more evident than in the fields of nutrition and health care. The prime providers of those packages are the agribusiness or food TNCs and the pharmaceutical TNCs.

Controlling food

Both food and good health are basic needs. From a TNC perspective, therefore, supplying those needs is merely a response to essential demands. Examining the food business, however, it becomes clear that a large measure of control enters the system, beginning with the seeds that are required, the purpose for which land is used, and runs through the processing, packaging and ultimate distribution of the products. Every step of the way there are large questions that need to be asked.

One of the foundations of today's agriculture is the belief that more food needs to be produced in order to feed the world. During the 1950s and early 1960s, scientists put their skills to this task and developed a series of new seeds which they called 'high yielding varieties' (HYVs). Concentrating mainly on staple crops—wheat, rice, maize—the HYVs were seen as an answer to the world's food problems. They were launched with a great flourish in the so-called 'Green Revolution' of the 1960s. The result was devastating.

First and foremost, the new seeds are not high yielding in and of themselves. They require correct soil conditions, specific types of irrigation and large amounts of fertilizers and chemical pesticides. They are more appropriately termed 'high responsive varieties' (HRVs). Without these inputs, which for the most part are expensive, farmers reported large-scale crop failures.[29]

This loss in crops almost inevitably leads to a cycle of debt and dependence, particularly for small-scale farmers in developing countries. The farmers are forced to borrow, usually at exorbitant rates of interest, using the next year's harvest as collateral. As more and more of the harvests go to pay off the interest, the only collateral left is the land. When that goes, the farmers have nothing, and either become low-wage employees on the larger farms or drift into the urban centres in search of alternative work.

Another danger is that, where previously a great diversity of crops and seed varieties had been used, now single crops are planted using single varieties of seeds. This transformation means that farmers, who had grown most of the food needed for their families and immediate communities and offered any surpluses for sale on the open market, are now forced to sell their entire crop in order to purchase the food they need for their own families.

Pat Mooney, who notes with concern that control over the seeds used is the first step in controlling the entire food system, comments:

> Both the crops and economies of Asia, Africa and Latin America have been hauled into the Western market economy under the pretext of feeding the hungry. The Third World is being brought into a food system which has not worked well in the First World, and which is in imminent danger of destroying alternative options for poor nations.[30]

Some very real and tragically ironic effects have occurred. While the Sahelian countries of Africa suffered a severe drought during the early 1970s which stimulated massive foreign aid to feed hungry populations, all but one of those countries produced and *exported* enough grain to feed their total population during the worst drought year.[31]

Brazil exports 97.5 per cent of its orange crop—much of it to the Coca Cola company, which in turn imports Fanta Orange soft drink to Brazil. Fanta, however, contains no orange juice and many Brazilians suffer from a vitamin C deficiency in their diets.[32]

The problem then is not so much the production of food, but what food is produced where, for what purpose, where and how it is ultimately distributed, and who can afford to buy it.

Defenders of the 'global farm' system of production point out that increased exports will help the foreign exchange earnings of countries, and therefore benefit the poor. That argument is incorrect for two reasons. First, the price of imports for developing countries has been increasing much more rapidly than the price of their exports. Bananas, for example, are the most important fresh fruit in international trade. For most of the countries in the Windward Islands in the Caribbean, bananas are the primary source of export earnings. One British-based TNC, Geest, buys the entire banana crop of the Windwards.[33] In 1960, 3 tons of bananas provided enough income to purchase a tractor. By 1970, 11 tons of bananas were required to buy the same tractor.[34] The second important point is that, like any of the factors in the 'trickle-down' theory of development, the export earnings simply do not reach the poorest sector of the community. In Senegal in 1974, 30 per cent of foreign exchange earnings were spent on refrigerators, air conditioners, alcohol, tobacco and similar imports for the elite urban population.[35]

The global farm system does have benefits for some, however—primarily the TNCs who supply the seeds, the fertilizers and pesticides, the elaborate machinery required to harvest the crops efficiently and cost-effectively, and those who process and distribute the produce.

To an ever-increasing extent, the TNCs are not simply brokers in an exchange between farmers and the general public. They control who grows what, with which seeds and which agricultural inputs, and provide the finance so that farmers will stay in line. At the other end of the scale, they help to manipulate the markets so the consumers will 'demand' the products a TNC can profitably supply.

This control operates in many countries, including the industrialized nations. Unilever, the world's largest food TNC, has exclusive contracts with over 1,000 farmers in the United Kingdom guaranteeing a constant supply of specific vegetables. A former chairman of Unilever happily admitted that the operations of the company had altered the agriculture of East Anglia in the United Kingdom. This, he said, provided much of the challenge to his job, 'the power to change things, the power not to have to accept things as they are'.[36]

The milk industry exemplifies that power to change. As James Basset, vice-president of foreign operations for the Borden company, explains, 'in a great many areas of the world, people don't know what milk is. We have to tell them what it is and where it comes from'.[37] While people in the dairy-rich countries of the world accept unquestioningly that milk is a necessary part of their diet, particularly for children, for much of the world's population milk is traditionally an unimportant luxury item.

However, with the over-production of milk in Europe, which guarantees cheap prices for raw milk or skimmed milk powder (SMP), it obviously makes commercial sense to find new outlets for the milk. One scheme in India, Operation Flood, looked like a golden opportunity. The theory of the scheme was to take SMP as food aid, reconstitute it into whole milk in purpose-built dairies and sell it through an effective distribution network. The proceeds would then be used to stimulate India's indigenous dairy industry and develop self-sufficiency in fresh milk.

Originally started in 1970 and intended to run until 1975, the first phase of the scheme was extended until 1979. By 1978, the number of litres of milk produced per day had indeed increased by two and a half times the 1970 figures. However, the volume of imported SMP had increased to three times its 1970 level.[38] The increase was such that food aid could no longer meet all the requirements and extra SMP had to be purchased on the open market. Two commentators have noted that this has provided an ideal opportunity for dairy manufacturers in the industrialized countries.[39]

Nestlé, as one of the leading milk companies despite not owning a single milk cow, has done very well out of the surplus SMP available on the world market. The company's 1978 annual report noted the 'considerable support' received from the EEC in the 'granting of export subsidies on dairy products'. In other parts of the world, Nestlé has encouraged farmers to switch from traditional crops to dairying, for example, in Brazil, Mexico, Peru and Colombia. However, when prices for SMP dropped on the world market in 1977, the company cut back on the purchase of whole milk, with the result that Latin American farmers had to throw milk away.[40]

With the raw materials ensured, both milk and food TNCs set about to maximize return on those products. The basic concept used is that increased profits will accrue when a product has 'added value'—that is, when the raw material is transformed into something more 'desirable' and therefore more expensive.

In Brazil, for example, price controls operate on whole milk. No price controls are in place for milk derivatives. It is therefore not surprising that up to 70 per cent of Brazil's whole milk is bought up by milk-derivative manufacturers, transformed into cheeses, yoghurts, milk-based cereals and milk desserts.[41] The result is that the consumer pays more for each gram of protein (or any other nutrient) in the product.

The example of what happens to a potato shows dramatically how much more the consumer pays for 'added value'. In 1979, loose potatoes in the United Kingdom sold for about 10p per pound. Dehydrated and powdered as Cadbury's Smash they cost 24p per pound. As frozen chips prepared by Unilever, they were 53p per pound. Potato crisps worked out at £1.30 per pound. Further adaptations producing snacks such as Smith's Frazzles cost over £2 per pound.[42] While there is no argument that the potato makes a useful nutritional contribution to the diet, there is little likelihood that crisps or similar snacks would ever be eaten in sufficient quantity to provide nutritional benefit. This brings up another important point. Potatoes are almost never advertised. All the other products are, often heavily. In fact, the most recent advertising for one brand of potato crisps makes no mention of the nutritional levels, but has as its selling point that they are the 'noisiest' crisps!

One of the major characteristics of food TNCs is that they rely heavily on advertising and marketing to push their products. In 1973, the British food industry spent the equivalent of £1.50 for every man, woman and child in the country, a total of £88 million, on the promotion of its products. This was more than 20 per cent of all advertising expenditure in the United Kingdom.[43] It is a pattern repeated around the world. In 1974, advertising for food accounted for 11.5 per cent of total advertising in Panama, 20.6 per cent in South Korea, 21.7 per cent in Taiwan, and 20.4 per cent in Austria.[44]

Lee Bickmore, chairman of Nabisco, explained the ambitious plans his company had for marketing Ritz crackers, certainly not a product that would rank very high on anyone's list of essential foods.

> Why, we plan someday to advertise all over the world. We might spend, say, $8 million for an advertisement on a communications satellite system. It might reach 359 million people. So what we are doing now is establishing the availability of our products in retail outlets all over the world.[45]

The purpose of heavy advertising and promotion expenditure is threefold:

- to expand the potential market for the generic product line by drawing attention to the availability and instil a desire for that range of products;
- to develop specific brand consciousness for a particular company's product, and increase market share;
- to erect barriers to entry into the market for competitors unable to match the promotion expenditure.

To develop a new market, a company with a new product, say, for example margarine, would examine the market and recognize butter as the chief competition. The first step would then be to launch an effective promotional campaign which highlights the advantages of margarine over butter, and announces Brand X as the margarine to buy. Once a niche has been established for margarines generally, the emphasis then shifts to promoting Brand X as the best, smoothest, purest, lightest, tastiest margarine, with the secondary message being its advantages over butter. From there on, once a market share is established, an occasional burst of advertising will suffice to keep Brand X in the minds of the consumer.

Replace infant formula for margarine, and breast milk for butter in this example, and a very real clue to the development of the infant formula market comes to light. In this way, the brand name becomes a household word. People think Coke when they mean cola; Nescafé when they mean instant coffee; Birds Eye when they mean beefburger; Uncle Ben's when they mean rice; Lactogen when they mean infant formula. Note that in the process, the idea of alternatives begins to disappear. Thirst becomes quenched by Coke or Nescafé, not fruit juice or water. Hunger is satisfied by Uncle Ben's or Birds Eye, not the unbranded, bulk-purchase rice from the market or the fresh minced beef from the butchers. Infants are fed with Lactogen, not breast milk.

The subtle message that underlies all such food advertising is that packaged, processed foods have special appeal or special benefits. They are portrayed as more glamorous, more exciting, more convenient, more reliable, and in some cases more nutritious.

While there are some limited controls in some countries that ensure that advertising does not lie, as Tudge points out, this does not mean that they tell the whole truth.[46] He uses the example of an advertisement announcing that frozen peas contain more vitamin C than tinned peas. It is a true statement, but needs a qualification such as: 'Yes, but fresh broccoli, now in season, contains more still; and peas in any case do not contribute significantly to your vitamin C balance; and the vitamin C in potatoes is cheaper'.

He recognizes that most food companies would resist including such qualifications and that any independent attempts to provide the qualifications would be expensive—an expense that would ultimately have to be borne by the consumer. One advertising expert in the United Kingdom argues that 'it would be impossible to control advertising if one couldn't start with the assumption that the role of advertising is to persuade the consumer and the fact that it consists of partial statements'.[47]

Tudge offers an idea that warrants some consideration. It, too, would generate vehement protest from food TNCs, for he suggests,

> It would be better not to advertise food at all. Indeed, since the most advertised products are the value-added ones, most liable to increase food bills, and requiring profligate use of energy and land, it is more pernicious to advertise food than to advertise cigarettes.[48]

Pushing pills

Despite a different approach in some cases, there are remarkable similarities between the food industry and the pharmaceutical industry. The first similarity is the control the industry exerts over the market. At first glance, such control would appear impossible, for the assumption would be that no company can control the incidence of disease and therefore cannot control which drugs are purchased when. Would it not be the case that drugs were only purchased when they were needed for specific treatments? The answer to the question is provided by a former medical director of E. R. Squibb & Sons: 'The incidence of disease cannot be manipulated and so increased sales volume must depend at least in part on the use of drugs unrelated to their real utility or need'.[49]

There are many ways of increasing drug sales: recommend their use for conditions for which they are ineffective or for which there is a more effective remedy; recommend increased dosages; move into new markets; make small changes and increase prices. For example, anabolic steroids, such as Sterling Drug's Winstrol Compound, should not be used to stimulate growth in children according to the American Medical Association. One researcher in the early 1970s found them being promoted for this use in Mexico, Brazil and the Dominican Republic.[50] Migril tablets for migraine relief, produced by Wellcome, carry a maximum recommended weekly dosage of 10 mg. in the United States, but 24 mg. in Africa and Asia.[51]

The control, however, begins at the research and development (R&D) stage. In terms of the diseases that could be effectively treated by medication, the vast majority of sufferers live in developing countries and are faced with a series of tropical diseases for which there are few, if any, drugs. Logic would suggest that priority research would be directed at those diseases. Commercial logic suggests otherwise. The former president of Parke–Davis was concerned about the research priorities of the company. 'Fifty per cent of the research at Parke–Davis during that time was focused on exotic Third World diseases. I steered it into more commercial applications'.[52]

This is not the policy of only one company. According to the British-based, industry-funded Office of Health Economics, the industry as a whole spends 1 per cent of its annual R&D budget on the specific health conditions of developing countries.[53] To put that figure into perspective, it should be remembered that two-thirds to three-quarters of the world's population live in developing countries. Another salient point is that the developing countries provide between 15 and 20 per cent of the pharmaceutical industry's annual sales turnover of some $100 billion. As Muller points out, research is largely funded from profits on sales—almost a company-imposed 'research tax'—so one could legitimately expect at least 15 per cent of the R&D budgets to go to tropical disease research.[54]

If the R&D went on signficant new therapeutic breakthroughs, there might conceivably be some justification, as eventually the widening-out of treatments could provide some preparations that were relevant to developing countries. However, what occurs is a 'deepening', not a widening, of research. Attempts are made to imitate or modify an existing, successful drug, so that a competing product can grab a share of the market. In 1980, only 119 of the 1,087 drugs presented to the United States Food and Drug Administration (FDA) for product licences were considered of either 'important' or 'moderate' therapeutic gain. The rest were 'me-too' drugs, imitative varieties which represented 'little or no' therapeutic gain.[55]

What that 'me-too' research means is a phenomenal number of drugs on the market: 14,681 in Brazil,[56] 36,000 in Switzerland,[57] 24,000 in West Germany and 21,000 in Italy.[58] Yet the World Health Organization has prepared a list of just over 200 essential drugs which are effective for virtually all the therapeutic needs of developing countries. Even allowing for some additional drugs for specific ailments, a list of 500–600 would cover the health needs in any country.

A PhD in economics is not required to recognize that such a limitation would generate a tremendous saving in health care costs. Nor would it require much thought to recognize that those savings could be addressed to dealing with improvements to social conditions such as clean water, sanitation and basic health and nutrition education which have had more impact over the years than all the drugs put together.

Thus, just as not all the products sold by the food TNCs are essential, neither are all drugs essential. Nowhere is this more evident than in the over-the-counter (OTC) trade—the non-prescription preparations which include a variety of cure-alls of dubious value.

One doctor estimates that one-third of all money spent on drugs world-wide is on these OTC preparations.[59] In Bangladesh, 25 per cent of drugs expenditure goes on vitamins, while in Nigeria 49 per cent of drug expenditure is on OTC preparations.[60] Even in the United Kingdom, a recent report by the respected consumers' magazine *Which*? concluded that the £45 million a year spent on vitamins was money largely wasted. The report advised, 'you should get all the vitamins you need from the food you eat'. According to one industry executive, the national press in the United Kingdom tried to play up the story, 'but found that most manufacturers did not dispute what *Which?* said'.[61]

It is in the OTC area that the pharmaceutical TNCs use promotion to the public in the same way that food TNCs promote value-added products. Glaxo, for example, pushes its Glaxose D—basically sugar—in public advertising in India as a 'treatment' for normal and natural tiredness that occurs after a hard day's work.[62] Sandoz, in Nigeria, uses small vans with loudspeakers to push a calcium and vitamin C tablet. Consumers are told the pills contain 'powerful energy you need'.[63] Sandoz explains its OTC advertising in this way:

> Specific claims are not made for any of these products, just vague statements promising power, long life, happiness, vitality, instant relief and contentment. The value of such promotion tactics may be questioned by the educated population but the uneducated Nigerian buys these products on the promises that are made and frequently continues to use them.

If the function of advertising is to push ideas, not just products, the main idea that is pushed is that drugs work. The attitude that is stimulated is that the only or best treatment for ill health is a drug. This has important implications when it is coupled with the vast amount of promotion the drug companies aim at health workers. In total, the pharmaceutical industry spends well over 10 per cent of sales turnover on promotion and advertising of its products—more than $10 billion per year, or just over $2 for every man, woman and child in the world.[64] That is well above the amount spent on R&D, as Table 5 shows.

The *Wall Street Journal*, in an examination of the way American Home Products achieved 46 per cent of the market in nitrate drugs for chest pains with Isordil, noted that it was 'promoted so aggressively that many doctors think only of Isordil when they think of nitrates'.[65] This was despite the fact that the United States' FDA had not approved all the claims the company made about the drug, and that some doctors found very little difference between it and competing drugs. The *Journal* said:

> If the success of Isordil illustrates anything, it's that promoting drugs to doctors these days is much like selling soap to consumers. It's all in the marketing. Indeed, the very survival of a drug in today's highly competitive marketplace often depends as much on a company's marketing talents as it does on the quality of its medicine.

Table 5 Profit margins, advertising and R&D expenditure of selected pharmaceutical companies as a percentage of total pharmaceutical sales, 1978

Company	Profit	Advertising	R&D
Abbott Labs.	20.0	11.5	7.5
American Home Products	24.0	10.0	3.3
Bristol–Myers	18.0	15.0	4.0
Sandoz/Wander	17.0	14.5	12.9
Syntex	25.0	15.0	10.0

Source: P. A. Borgoltz, Ch. 12, in D. B. Jelliffe and E. F. P. Jelliffe (eds), *Advances in International Maternal-Child Health*, Vol. II, Oxford, Oxford University Press, 1982, p. 164.

A senior advertising copywriter, Paul Garai, confirmed this approach when he noted: 'razzle-dazzle promotion has come to play a more decisive role than scientific evaluation of drugs in determining the career of pharmaceutical products.'[66]

In 1980, American Home Products was the sixteenth-largest advertiser in the United States with an expenditure of $197 million, although it was fourth in television time bought, coming behind Proctor and Gamble, General Foods, and Ford.[67] In 1974, pharmaceutical companies accounted for 26 per cent of all advertising in Taiwan, 18.6 per cent in South Korea, 10.4 per cent in Austria, and 7.7 per cent in Panama.[68]

As with food TNCs, the function of the advertising is to stimulate brand consciousness. This has obvious benefits to the TNC, but costs the consumer and the national economy. In most cases, generic drugs—drugs sold under the name of the active ingredient—are as effective as the branded drugs and are available at a much lower price. In the United Kingdom and Italy, for example, the generic equivalent to Roche's Librium is available at prices 25 per cent and 30 per cent lower than Roche's product, yet the generic producers have been unable to capture more than 3 per cent or 4 per cent of the market, thanks to the high-pressure promotion by Roche.[69]

Selling prescription drugs is an ideal situation for a pharmaceutical company. It has a captive channel to the market—the health worker—who can be targeted easily, who is not particularly concerned about the cost of the product, since the ultimate cost will be paid by the patient (or an insurance plan or the government) and, perhaps most importantly, the company is liable to be the only source of information about a particular drug. Like nutrition, pharmacology does not figure large in the training programme of most doctors.

Professor Uvnas, head of the Pharmacology Department of the Karolinska Institute in Stockholm, commented in 1968 that doctors do not use their drugs with optimal efficiency and wisdom. One important reason, he said, was the poor quality of the information, not the lack of quantity. On one day, Swedish physicians at a hospital attached to the Institute received a total of 5,775 items of direct mail advertising from the pharmaceutical industry.[70]

The purpose of this vast amount of information clearly is not to inform. Abbott Laboratories describes a medical magazine, *Abbott Topics*, which it mails four times a year to 250,000 readers in the United States, including doctors, interns, medical students and hospital personnel, in this way:

> As the voice of Abbott, *Abbott Topics* can be a positive force in moulding the physician's opinion of Abbott. The implications of having the physician think and speak well of Abbott, and engender the same attitude among his fellow MDs, among the druggists, and even the public, are incalculable. In effect, we are striving to make the physician a low-pressure salesman of Abbott.[71]

Promotion does not stop with leaflets, direct mail flyers or magazines, however. As Dr Vernon Coleman notes, drug companies entertain lavishly. 'As a practising general practitioner I have found that it would be possible to attend at least one luncheon or dinner every week and probably more than one.'[72] He also comments that free gifts, samples and support for travel expenses at meetings and seminars are commonplace. It is a pattern that is repeated throughout the world. A hospital administrator in the Philippines told one researcher, 'It's so corrupting to be aware that you could have almost anything you asked for'.[73]

The promotion not only sells drugs, but it works at the level of buying the names of doctors to support dubious research findings. A British doctor comments:

> To my surprise a respected drug representative recently offered me money and a free trip to the Continent, in return for joining a drug trial organised by his company. The 'trial' seemed to me to be designed to provide commercially useful data rather than medical information.[74]

All of this adds up to high costs, often ineffective (and sometimes dangerous) treatment, and a focus of attention on symptoms, not causes. In a world where resources, both human and material, are finite and over-stretched, this approach to health care is, at best, wasteful.

Both the pharmaceutical and food manufacturers, for all their glowing rhetoric about service to society are largely missing the point. This is not to deny the valuable benefits that have been derived from the considerable investment pharmaceutical TNCs have made on some very important and very useful drugs. Nor is it to deny the very real contribution that food technology has made to better storage procedures, some improvements in distribution systems, and some increased food production.

It is not a black-and-white situation where all the activities of TNCs are wrong, or where they are simply profiteering, uncaring businesses. The charge that needs to be laid at their doors is not so much irresponsibility as *irrelevancy*. The bulk of their research, their products and the promotion and marketing of those products is irrelevant to the basic needs of the mass of the people who inhabit this planet. It is their failure to come to terms with this irrelevancy, and to use the technology and skills they have available to integrate a new dimension into their activities which generates the charge of irresponsibility.

The point is ably put by the president of General Foods (GFC):

> What is it that GFC can contribute to a foreign subsidiary? Well, first we have more than 10% of all the food researchers in private industry in this country, and therefore we have a capability in food technology to contribute. Our Dream Whip and Gainsburger dog food products, for example, were technical achievements.[75]

Undoubtedly they were. But artificial cream and better dog food are nothing to boast about in a world where one-quarter of the population go to sleep hungry, wondering if there will be any food the next day.

The issue which has confronted this irrelevancy most strikingly and which has exposed both food and drugs TNCs to criticism is the marketing of artificial infant feeding—both the concepts and the products. As Tudge notes, it illustrates the wide gap between the stated intention of companies and the actual consequences of their actions; how all-consuming the commercial imperative can be; how easily people can be persuaded to go against their own best health interests; and how quickly traditions which protect those interests can be changed.[76] It also shows that more and more people are refusing to accept this irrelevancy on the part of TNCs.

Notes

1. P. A. Borgoltz, 'Economic and business aspects of infant formula promotion: implications for health professionals', in D. B. Jelliffe and E. F. P. Jelliffe (eds), *Advances in International Maternal-Child Health*, Vol II, Oxford, Oxford University Press, 1982, p. 163.
2. Department of Health and Social Security (UK), *Present Day Practice in Infant Feeding 1980*, London, 1980, p. 15.
3. UN Centre on Transnational Corporations (CTC), *Transnational Corporations in Advertising*, document ST/CTC/8, New York, 1979, p. 9.
4. N. Janus and R. Roncagliolo, 'Advertising, mass media and dependency', *Development Dialogue*, 1, 1979, p. 84.
5. UN CTC, op. cit., pp. 34–5.
6. Janus and Roncagliolo, op. cit., p. 82.
7. K. P. Sauvant and F. G. Lavipour (eds), *Controlling Multinational Enterprises: Problems, Strategies, Counterstrategies*, Boulder, Colorado, Westview Press, 1976; cited in UN CTC, op. cit., p. 34.

8. Ibid.
9. A. Toffler, *The Third Wave*, London, Pan Books, 1981, p. 56.
10. Janus and Roncagliolo, op. cit., p. 95.
11. C. Medawar, *Insult on Injury?*, London, Social Audit, 1979, p. 12.
12. Ibid., p. 11.
13. M. C. Thordahl, 'National and multinational opportunities for the encouragement of innovation', *Food Technology*, January 1982, p. 102.
14. Nestlé, *Nestlé in the Developing Countries*, Vevey, 1975, front cover.
15. S. George, 'Nestlé Alimentana SA: the limits to public relations', *Economic and Political Weekly* (India), **XIII**, 37, 16 September 1978, pp. 1591–1602.
16. Medawar, op. cit., pp. 14–15.
17. W. Brandt, *et al.*, *North-South: A Programme for Survival–Report of the Independent Commission on International Development Issues*, London, Pan, 1980, p. 187.
18. Janus and Roncagliolo, op. cit., p. 96.
19. S. Allende, 'Speech to the United Nations', in H. Radice, (ed.), *International Firms and Modern Imperialism*, London, Penguin, 1975, pp. 239–40.
20. R. J. Ledogar, *Hungry for Profits*, New York, IDOC, 1975, p. xi.
21. E. Gabus, 'Multinational companies: the external relations', in G. Curzon and V. Curzon (eds), *The Multinational Enterprise in a Hostile World*, London, Macmillan, 1977, p. 130.
22. M. Muller, *The Health of Nations*, London, Faber & Faber, 1982, p. 194. Pfizer, not surprisingly, argues a different case on the Sri Lanka question and over its role in developing countries in general. Pfizer told Oxfam that 'the facts of the case to which you refer are substantially different from what we have recorded . . . We acted quickly, persisted in our efforts to help, and were responsible, with reference to price and other matters of detail that developed in this situation'. Pfizer then goes on to put the blame on the Sri Lankans and their state pharmaceuticals agency which was unable 'to cope adequately with the health needs of the Sri Lankan people'. For further detail, see D. Melrose, *Bitter Pills*, Oxford, Oxfam, 1982, p. 187.
23. Muller, op. cit., p. 196.
24. A. Sampson, *The Seven Sisters*, New York, Bantam, 1976, p. 374.
25. Medawar, op. cit., p. 11.
26. Ibid., p. 12.
27. J. K. Galbraith, 'The uses of an economic system—and of economics', in M. Green and R. Massie Jr. (eds), *The Big Business Reader*, New York, The Pilgrim Press, 1980, p. 479.
28. I. Illich, 'Outwitting the "developed" countries', in H. Bernstein (ed.), *Underdevelopment and Development*, London, Penguin, 2nd edn., 1976, p. 358.
29. P. R. Mooney, *Seeds of the Earth*, London, ICDA, 1980, p. 43.
30. Ibid., pp. 43–4.
31. F. M. Lappe and J. Collins, *Food First*, London, Souvenir Press, 1980, p. 78.
32. Ledogar, op. cit., p. 114.
33. War on Want, *Fruits of Development*, London, 1980.
34. Lappe and Collins, op. cit., p. 155.
35. Ibid., p. 77.
36. C. Tudge, *The Famine Business*, London, Faber & Faber, 1977, p. 81.
37. *Printers' Ink* (US), 'Powdered milk to African tribesmen: Borden's world-wide marketing mission', 13 June 1958.
38. B. Dogra, *The Milk Muddle*, New Delhi, Social Change Papers, 1980, p. 11.
39. Ibid., p. 34. See also J. Clark, *Milking Whom? A Study of Europe's Leading Agricultural Sector, and its Effects on European and Third World Food Systems*, London, ICDA, 1979, p. 48.

40. Clark, ibid., p. 64.
41. C. Monteiro, *The Harmful Activity of the Multinational Food-Producing Corporations on Child Nutrition in Brazil Through its Destructive Effects on the Organisation of the Production of Pasteurised Milk for Immediate Consumption*, mimeo paper prepared for workshop on 'Rethinking Infant Nutrition Policies under Changing Socio-Economic Conditions', sponsored by the International Union of Nutritional Sciences, Oslo, Norway, 25 November–5 December 1980, pp. 11–13.
42. Lappe and Collins, op. cit., p. 230. (In the United States, chips are known as French fries and crisps are known as potato chips.)
43. Ibid., p. 231–2.
44. R. Roncagliolo and N. Janus, 'Advertising and the democratisation of communications', *Development Dialogue*, 2, 1981, p. 37.
45. Lappe and Collins, op. cit., p. 237.
46. Tudge, op. cit., p. 112.
47. M. McComas, *Europe's Consumer Movement: Key Issues and Corporate Response*, Geneva, Business International, 1980, Part II, p. 86.
48. Tudge, op. cit., p. 113.
49. Ledogar, op. cit., p. 25.
50. Ibid., pp. 28–9. (A recent study by the International Organization of Consumers' Unions, *Anabolic Steroids: Availability and Marketing*, August 1983, showed that the practice was still prevalent in the 1980s. The study found only two indications for the use of Winstrol in the United States: aplastic anaemia, and as 'probably' effective for the treatment of senile osteoporosis. 'The US FDA noted that in the latter case, anabolic steroids are without value as primary therapy'. However, in Malaysia, Winstrol was indicated for ten conditions, including trauma, fractures, severe burns, kidney disease, convalescence from acute illnesses, and malnutrition due to alcoholism. In the Philippines, Indonesia and West Germany, other anabolic steroids produced by Winthrop were found with indications for use in treating malnutrition, weight loss or poor weight gain in children. Fourteen other anabolics manufactured by other companies were also found being promoted for use in children in India, Indonesia, Malaysia, the Philippines and West Germany.)
51. Medawar, op. cit., p. 117.
52. Muller, op. cit., p. 62.
53. D. Melrose, *Bitter Pills*, Oxford, Oxfam, 1982, p. 53.
54. Muller, op. cit., p. 66.
55. Ibid., p. 67.
56. Ibid., p. 138.
57. Ibid., p. 135.
58. Mooney, op. cit., p. 95.
59. V. Coleman, *The Medicine Men*, London, Arrow Books, 1977, p. 89.
60. Muller, op. cit., pp. 99, 92.
61. Consumers' Association, 'How to buy vitamins', *Which?*, January 1984, p. 6. See also P. Tisdall, 'Medicinal maelstrom', *Marketing*, 12 April 1984, p. 28.
62. Medawar, op. cit., p. 85.
63. Muller, op. cit., p. 92.
64. Ibid,. p. 116. (Figures adjusted to 1985 market levels.)
65. M. Waldholz, 'Pill promoters: marketing often is the key to success of prescription drugs', *Wall Street Journal*, 28 December 1981.
66. Coleman, op. cit., p. 94.
67. Waldholz, op. cit.
68. Roncagliolo and Janus, op. cit., p. 37.

69. S. Lall, *The Multinational Corporation*, London, Macmillan, 1980, p. 171.
70. Coleman, op. cit., pp. 92, 105.
71. *Direct Mail and Mail Order Handbook*, 1981, cited in *INFACT Update*, November 1981, p. 4.
72. Coleman, op. cit., p. 101.
73. Muller, op. cit., p. 75.
74. M. Weitz, *Health Shock*, London, Hamlyn Paperbacks, 1982, p. 157.
75. Lappe and Collins, op. cit., p. 239.
76. Tudge, op. cit., p. 110.

3. Public pressure

> If she continues to feed the baby with a dirty bottle and dirty water, it can lead to the death of the child. We accept no responsibility for hygienic conditions in the country and for the lack of knowledge of writing and reading. . .
>
> Dr Hans Rudolf Muller, vice-president, Nestlé, in BBC TV programme, *Panorama*, 1 December 1975

Dr Cecily Williams made medical history. In the early 1920s she became the first white woman doctor to work in Africa. She concentrated on paediatrics and was the first doctor to identify and name the severe protein deficiency condition in young children, *kwashiorkor*. It is a West African word which literally means a child displaced or deposed from the breast.

Dr Williams noted that the onset of the condition was associated with the time of weaning from the breast when high-carbohydrate, but low-protein, foods were introduced into the child's diet. It usually occurred during the second year of the child's life and was often accompanied by other infections, diarrhoea and all too often led to death.

Several years later, while working in Singapore, she noted that malnutrition was striking at an earlier age, and in a different form: marasmus, which is both energy and protein deficiency. She also noticed that this malnutrition was closely related to artificial infant feeding. In a hard-hitting speech to the Singapore Rotary Club in 1939 entitled 'Milk and Murder', she condemned the spread of artificial infant feeding.

> It is your fault, it is public opinion, misguided and obstinate as it is, that is murdering so many children . . . If you are legal purists you may wish me to change the title of this address to 'Milk and Manslaughter'. But if your lives were embittered as mine is, seeing day after day this massacre of the innocents by unsuitable feeding, then I believe you would feel as I do that misguided propaganda on infant feeding should be punished as the most criminal form of sedition, and that these deaths should be regarded as murder.

The main product used for artificial infant feeding in Singapore in those days was sweetened condensed milk (SCM)—a product which is totally unsuitable for infants. In 1911, following forty years of campaigning by doctors, the British government required that all tins of SCM should be labelled with a clear warning that it was not to be used for feeding infants.[1]

Dr Williams did not specifically implicate the infant food manufacturers in her Singapore speech, although they were involved in the 'misguided propaganda'. In 1936, for example, a diary for doctors, prepared by Nestlé and circulated in South East Asia, contained two full pages with glowing praise for the benefits of SCM.

> Sweetened condensed milk is the food *par excellence* for delicate infants . . . We have not employed Nestlé's Sweetened Condensed Milk over long periods, but we do know that it gives excellent results in the hand feeding of young infants.[2]

Some years later, Dr Williams was much more scathing about company promotion—in particular the activity of 'milk nurses', the sales representatives the companies used to promote their products to mothers. 'Then I went to Singapore and found Nestlé nurses, these girls dressed as nurses, dragging a good lactating breast out of the baby's mouth and pouring in baby milks.'[3]

Although the evidence linking malnutrition with artificial infant feeding began to grow in the years following the Second World War, it was not until 1968 that the link between artificial infant feeding and commercial promotion was authoritatively raised. Once again, the message came in a speech. The speaker was Dr Derrick Jelliffe, who at the time was Director of the Caribbean Food and Nutrition Institute in Jamaica. 'It is harsh, but correct, to consider some of these children as suffering from "commerciogenic malnutrition"—that is caused by the thoughtless promotion of these milks and infant foods.'[4] He called for a dialogue between health professionals and the infant food industry to try to resolve the problem. Dr Catherine Wennen followed up this call with an article in *Tropical and Geographical Medicine* which drew attention to the practices of the milk companies.[5]

Dr Wennen worked in Nigeria from 1963 to 1967 and was shocked by the high level of promotion for infant formulas. One of the radio advertisements she heard broadcast regularly was: 'Mother, believe in Lactogen . . . All things in mother's milk are also present in Lactogen. Mother, watch the health of your baby, and give him the best, give Lactogen'.[6]

Dr Wennen approached the Nestlé manager in Lagos 'in order to point out the sad consequences of their indiscriminate sales promotion. He ignored our criticisms and arguments. It looked as if he did not see our point, but obviously he did not like to enter into a dialogue'. She noted that this was during a time when the annual meeting of the Society for Health of Nigeria had suggested that a 'relentless campaign against bottle feeding should be launched'. Her main conclusion about industry promotion was: 'A lot of damage has been done and it still goes on.'[7]

The UN becomes involved

Dialogue was only going to be successful if it involved as many companies as possible. With this in mind, Jelliffe and another eminent paediatrician, Dr Bo Vahlquist, asked the UN Protein-Calorie Advisory Group's (PAG) *ad hoc* Working Group on Feeding the Pre-School Child to organize a meeting. It was held in Bogotá, Colombia in November 1970 and included twelve paediatricians, twelve representatives from the industry, plus representatives from several UN agencies including UNICEF and the Food and Agriculture Organization (FAO).

The meeting was held behind closed doors in order to ensure full industry participation and no comprehensive report of the proceedings was ever made public.[8] One of the participants, David Cox, former president of Ross Laboratories, noted that the meeting was the first occasion when concerns about industry practices were brought to his attention, and he also noted that the meeting did not agree on all points.[9] The main disagreement was over the impact of the availability and promotion of infant formula on the decline of breast feeding.

Nevertheless, some tentative agreement was reached, and the discussion continued at a series of further meetings around the world. A workshop at the International Congress of the International Paediatrics Association in Vienna in August 1971 had many of the Bogotá participants present, and was followed by a PAG meeting in Paris in June 1972 which led to the preparation of 'PAG Statement No. 23'. It was subsequently revised in November 1973. The PAG statement was a godsend for the industry, for although it made reference to the benefits of breast feeding, it read like a catalogue on how to improve infant formula sales.

> In any country lacking breast milk substitutes, it is urgent that infant formulas be developed and introduced ...
> Governments should provide fiscal and financial incentives that will encourage industrial investment ... consider subsidy programmes to provide nutritious infant formulas ... provide also for free distribution of these foods to the vulnerable groups in the lowest income families ... provide ... practical training in correct product usage for all mothers ... stimulate use of mass media channels for both educational and ethical product promotion communication ... encourage recognition of promotion as an essential component of any marketing programme to establish widespread consumer use of nutritious foods for children. For approved products, the promotion of which is considered desirable in specific locations from a public health standpoint, governments should give active support to promotion programmes, including, where appropriate, financial assistance ...[10]

It was the first attempt by the industry to get an official stamp of approval for both the products and the marketing practices. Better than that, it called for the further entrenchment of industry influence with governments to the point where, in some cases, the governments themselves would handle the expense of distributing the products, or even pay for the advertising of the products.

It was a perfect double message: breast milk is best; infant formula is virtually the same. It destroyed the possibility of a coherent approach to infant feeding that Elisabet Helsing called for in 1981:

> Ministries of social affairs, supported by the health profession as well as by mothers' groups, tell mothers that breast feeding is the nutritionally and

> psychologically advantageous thing to do. On the other hand, the baby food industry, supported by health professionals, tells mothers that if they prefer, artificial feeding is a perfectly acceptable alternative. This double message creates a certain confusion in the mothers, a confusion which I myself have had ample opportunity to observe.[11]

The PAG statement had only two points to make about promotion and marketing: anything that 'could discourage appropriate breast feeding' should be avoided; and that promotion to mothers in hospital nurseries just after birth was not appropriate. It was a naïve document and the infant food industry had won the first round, although at the time, it looked as if it would be the only round. No one else was paying much attention to the problem. The health workers could be won over eventually. The UN system could be seen to have done something, and the industry was quite prepared to go on talking and collaborating with the UN and the health professionals as long as there was no major change proposed to limit the expansion of the baby milk market.

One company, Abbott/Ross, developed a 'broad policy' as a result of the meetings and the statement. In their words, it included 'enthusiastic endorsement of breast feeding as the *formula* of choice in developing countries' (emphasis added). It also called for 'restrained promotional practices, and especially the elimination of direct consumer promotion'.[12] As a pharmaceutical company with strong links with the medical profession, and with a small share of the market in developing countries, but a large slice of the American market, it was an easy concession to make.

In another part of the UN system, an attempt by the International Organization of Consumers' Unions (IOCU) to introduce some stronger controls on advertising and marketing of infant milks made little progress. IOCU prepared a memorandum on the issue and a draft 'code of practice for advertising of infant foods' at the end of 1971 and submitted it to the FAO/WHO Codex Alimentarious Commission early in 1972.[13] The Codex Commission is the UN body which deals with international quality and labelling requirements of products meant for human consumption. IOCU, by virtue of its NGO status with the UN Economic and Social Council (ECOSOC), was represented on various Codex committees on standards.

The draft code, while not calling for a total ban on advertising, did call for some restrictions in content, suggested that no product should be advertised for infants under three months of age, and called for an end to the distribution of samples of the products in health care facilities. It was a reasonable working document. However, the Codex Commission felt it was outside its area of competence and so it received little discussion.

The issue becomes public

Within a twelve-month period from the end of 1973 until the beginning of 1974, the issue moved from quiet, secluded discussion at UN meetings and exploded on to the pages of the world's press. In August 1973, the *New Internationalist* magazine in the United Kingdom ran a cover story feature based on interviews with Dr R. G. Hendrickse and Dr David Morley, two child health specialists with long experience in developing countries.[14]

Both doctors were adamant that commercial promotion played a role in the decline of breast feeding and the increased malnutrition of infants. Hendrickse commented that the 'vigorous advertising' had created 'serious health problems'. Morley noted that although the industry would like people to believe that it was behaving responsibly, 'if you read their advertisements the weight is still on bottle feeding'.

The *New Internationalist* called for action in a campaigning editorial: 'We must not accept the creation of more malnutrition for the sake of commercial gain by the selling of artificial baby milk to mothers who do not need it, cannot afford it, and are not able to safely use it'. Prior to publication of the story, the *New Internationalist* approached several organizations in the United Kingdom asking them to assist in the development of a public campaign to press for changes in company practices. War on Want, a London-based development agency, already aware of some of the situation from its support for health care programmes in developing countries, responded with interest, but wanted a stronger case. It commissioned freelance journalist Mike Muller to investigate further.

Muller turned out a devastating report, *The Baby Killer*, which War on Want published in March 1974.[15] Concentrating largely on the situation in Africa, the report drew together medical and promotional evidence in a comprehensive summary of the problems related to infant feeding. It attracted widespread media attention, prompted questions in the British parliament and brought the infant food industry into the public spotlight. Requests for copies, letters of praise and requests for permission to translate the report flooded in to War on Want, which immediately began to organize campaigning activities around the United Kingdom. In all, more than 20,000 copies of the English language edition of the report were sold or distributed world-wide. It was translated into German, French, Spanish, Portuguese, Norwegian, Dutch and several other languages by NGOs around the world.

The two companies which were most strongly criticized in the report were British-based Cow & Gate and Swiss-based Nestlé. Cow & Gate responded quickly and sent a team of senior executives to Africa and Asia to investigate. The company formally offered to withdraw its products from sale and distribution, but were told by government authorities and paediatric experts that that was not the solution, but that 'better techniques of education and better controls on promotional activities were required'.[16]

The managing director of Nestlé, Arthur Furer, expressed similar comments, noting that marketing needed to be 'handled with great care in order to avoid misuse of these important food products'. He added, 'I am personally glad to have read Mike Muller's report which gave me an opportunity to rethink the whole problem'.[17]

Libel case

Nestlé had very little time to 'rethink', however, before the German translation of the report appeared in May 1974. Translated by the small Arbeitsgruppe Dritte Welt (AgDW—Third World Action Group) in Berne, Switzerland, the introduction was rewritten and it was given a new title: *Nestlé Toetet Babys* (Nestlé Kills Babies). The Swiss press gave it wide coverage, and Nestlé responded by filing libel charges against the group in July 1974. Nestlé cited four specific counts:

- the title;
- the charge that the practices of Nestlé and other companies were unethical and immoral (in both the original report and the rewritten introduction);
- the accusation of being responsible for the death or the permanent physical and mental damage of thousands of babies by its sales promotion policy (in the rewritten introduction);
- the accusation that in the developing countries, the sales representatives for baby food are dressed like nurses to give the sales promotion a scientific appearance (in the original report and the translation).[18]

Nestlé offered to settle the matter quickly, and out of court, provided that the AgDW:

- apologized publicly and withdrew the statements against Nestlé;
- paid an amount to be determined by the judge to a charity;
- paid all legal costs;
- destroyed all copies of the report;
- guaranteed that they would not spread further accusations against the company.

The AgDW, claiming such conditions were only intended to silence their criticism, refused to accept. 'We prefer the trial and we will evoke evidence of the truth . . . The trial about the consequences of processed baby food in developing countries will be of public interest.'[19]

The trial went through three hearings: November 1975, and February and June 1976. The AgDW amassed comprehensive dossiers for each of the hearings, including testimony from health workers all over the world. Just before the final hearing, Nestlé withdrew three of the charges, leaving only the complaint about the title. The judge found the thirty members of the AgDW guilty of libel. In explaining the verdict, the judge said,

> The adequate causal connection between the sale or any other type of distribution of powdered milk and the death of infants fed with such products is interrupted by the action of third parties, for which the complainant, in terms of criminal law, cannot be held responsible. In this sense, there is no negligent or even intentional killing.[20]

The members of the group were fined 300 Swiss francs (US$150) each.

Nestlé, however, did not get a clean bill of health. The judge noted that the company's advertising in developing countries 'went considerably further' than in industrialized countries and that many of the statements made about breast feeding in Nestlé material were 'in no way appropriate for mothers in developing countries'.[21] The judge went on:

> It is considered proved by the evidence that the Nestlé company uses health nurses who have an advertising task and who, by their activity, have an advertising effect ... Hence, the need ensues for the Nestlé company fundamentally to rethink its advertising practices in developing countries concerning bottle feeding, for its advertising practice up to now can transform a life-saving product into one that is dangerous and life-destroying. If the complainant in future wants to be spared the accusation of immoral and unethical conduct, he will have to change advertising practices.

Nestlé while pleased with the decision on the libel count, refused to acknowledge that its practices were immoral or unethical. In a circular to Nestlé staff a few days after the judge's decision, Furer said that he had personally investigated the marketing practices of the company. 'I was able to see that they were normal and usual advertising methods, used by manufacturers of such products all over the world ... We must affirm that we have full confidence in the ethical basis of our actions.'[22] It was that continued affirmation of ethical behaviour that was to maintain Nestlé as the principal target of the pressure campaign.

The Nestlé trial generated a great deal of public attention and massive media coverage. Perhaps more importantly, it provided the first clear organizing focus for the entire campaign, both for the AgDW and for the growing number of other groups and individuals who were concerned about the issue. It stimulated the collection of additional data and began to forge the international links that would prove so important to the campaign in later years.

In July 1976, a three-day meeting was held in Berne which for the first time brought together individuals and groups from eight countries to share information and experiences and discuss future strategy.[23] The main outcome of the meeting was an agreement to start a loose information network, with one person from each country responsible for circulation of information on the issue to other countries. It was also agreed that a small fund be established to provide basic travel expenses to allow for the collection of information about industry practices in developing countries. Although the participants in the meeting were from industrialized countries in Europe and North America, they recognized the need to broaden their contacts with organizations and individuals in developing countries and agreed that this should be a priority activity.

The experience of the AgDW having been singled out by Nestlé in the hopes of squashing opposition and criticism provided important lessons for the NGOs involved in the campaign. If the campaign was to bring about improvements in infant feeding and restrict unnecessary marketing practices, there needed to be an international effort with the various groups working together. They could not afford to be isolated.

During the early 1970s, isolating the critics was the prime strategy for TNCs. When the UN established a Committee of Eminent Persons to study the impact of TNCs, six Swiss TNCs, including Nestlé, formed a small strategy group to infiltrate the UN committee and ensure that its findings were not all that critical.[24]

At the same time, the TNCs devised a basic five-point strategy for dealing with critics.[25] First, the critic was to be *discredited* as a discussion partner by being identified *as an opponent to the system*. Coupled with this was a general defence which proclaimed that *defending free enterprise was in the public interest*. TNCs were to be seen as these defenders and thus working in the public interest; the critics, therefore, were enemies of the people.

Following on from this, *dubious motives* were to be ascribed to the critics—ideological or nationalistic prejudices, ignorance or a lack of expertise were to be played up. Fourthly, criticism was to be disproved by the use of *isolated instances*—using the exception as the rule. Thus if a company was criticized for widespread malpractices, an example of good behaviour was to be publicly paraded. In conjunction with this, where criticism was indisputable, it was always to be referred to as an *individual case*, and preferably one that was being carefully investigated by the company.

Many of these strategies were used by Nestlé during the libel trial, even to the extent that the company hired a private investigation bureau in Zurich, which specialized in investigating alleged 'subversives', to spend four days gathering information on the thirteen members of the AgDW.[26] The judge in the Nestlé trial was not impressed by attempts to discredit the members of the AgDW. Noting that they were all in 'good repute', he commented:

> They are people with a generally high level of education who put a high ethical value on the work they do in development research and information and who want to be taken correspondingly seriously ... It is inadmissible to totally dismiss the accused as leftist revolutionaries.[27]

Dealing with the strategy developed by the TNCs demanded a broad base of support, constantly updated information, and a careful analysis which showed that the events were not isolated incidents, but part of an integrated world-wide pattern of marketing. The evidence gathered for the Nestlé trial provided a strong foundation, as did the testimony of respected health professionals. But the momentum of the campaign was hard to sustain without the specific focus which the trial had provided. A new focus needed to be found.

The TNCs, meanwhile, were not slow to recognize that they too needed to organize internationally. Nestlé had carried most of the criticism in the early years of the campaign, but an increasing amount was rubbing off on the other companies. They recognized the dangers of being picked off one by one and began building a buffer organization to deflect the growing criticism.

Notes

1. A. Chetley, *The Baby Killer Scandal*, London, War on Want, 1979, p. 149.
2. Consumers Association of Penang, *The Other Baby Killer*, Penang, 1981, p. 17.
3. C. D. Williams, interivew in the *Lansing Star*, 18 October 1978.
4. D. B. Jelliffe, 'Commerciogenic malnutrition?', *Food Technology*, **25**, 55, 1971.
5. C. A. M. Wennen-van der May, 'The decline of breast feeding in Nigeria', *Tropical and Geographical Medicine*, 21, 1969, pp. 93–6.
6. Swiss Information Groups for Development Policy (SAFEP), *Does Nestlé Kill Babies?*, mimeo, 1975, p. 8.
7. Ibid., p. 10.
8. Personal correspondence with one of the organizers of the meeting.
9. US Senate, *Marketing and Promotion of Infant Formula in Developing Countries: Report of the US Senate Health and Scientific Research Sub-committee*, Washington, US Government Printing Office, 1978, p. 198.
10. Protein–Calorie Advisory Group of the United Nations, 'Promotion of special foods (infant formula and processed protein foods) for vulnerable groups', *PAG Statement, No. 23*, 18 July 1972, revised 28 November 1973.
11. E. Helsing, 'Infant feeding practices in Northern Europe', in *Assignment Children*, 55/56, Geneva, UNICEF, 1981, pp. 85–6.
12. US Senate, op. cit., p. 269.
13. International Organization of Consumers' Unions (IOCU), *Draft Code of Practice for Advertising of Infant Foods*, The Hague, December 1971.
14. H. Geach, 'The Baby Food Tragedy', *New Internationalist*, August 1973.
15. M. Muller, *The Baby Killer*, London, War on Want, 3rd edn. 1977.
16. Personal correspondence from Ian Barter, chairman of Cow & Gate's international division, 7 August 1979.
17. Personal correspondence from Arthur Furer, chief executive officer of Nestlé, 8 August 1979.
18. A. Chetley, op. cit., p. 109.
19. SAFEP, op. cit., p. 7.
20. SAFEP, *Nestlé Case Will Not Continue–But Dispute Goes On*, information for the press, No. 5, mimeo, December 1976, p. A5. (This is an unofficial translation into English from the German done by AgDW.)
21. Ibid., p. A4.
22. Ibid., p. B3.
23. CCPD and ISIS, *Report on Baby Foods Meeting*, mimeo, Geneva, July 1976.
24. R. Edwards and S. Newens, *The Multinationals*, London, Liberation, 1979, pp. 12–16.
25. Ibid., p. 24.
26. SAFEP, (December 1976), op. cit., Appendix C.
27. Ibid., pp. A6–7.

4. The new protagonists

> For eight years, the industry's critics have borne the burden of proving that commercial marketing practices were actually contributing to infant malnutrition and morbidity ... From this time forward, the sellers of infant formula must bear the burden of proving that their products are not only desired, but actually serve a public purpose in the developing world.
>
> Dr James E. Post, Professor of Management Policy, Boston University, 23 May 1978[1]

One of the ideas that emerged from the discussions sponsored by the UN PAG was the establishment of regional and national industry councils to act as a forum for continuing discussions of the problems associated with infant feeding. At the Singapore seminar in November 1974, the PAG presented a paper outlining how such councils could be set up and what their function would be.[2]

It noted that there was a need to coordinate the efforts of *all* groups and individuals concerned with infant and child nutrition. In its list of groups, however, it failed to mention the need to involve the most important groups—women's organizations, consumer groups and mothers. Instead, it suggested councils composed of health workers, industry and government officials. The infant food industry was expected to put up the funds for these councils. Their basic purpose would be to discuss the most relevant findings on infant feeding and help implement some of the recommendations which had emerged from the series of PAG seminars. Also included was the possibility of assessing the practices of member companies if they went against agreed decisions.

To a large extent, it was an unworkable proposal. However, it did spawn an idea. Industry representatives at the Singapore seminar began to frame a different type of council. In early 1975, industry representatives met in Zurich to prepare for the establishment of an international council, exclusively composed of infant food companies. In November 1975, two days before the start of the first hearing of the Nestlé trial, the formation of the council was announced. The International Council of Infant Food Industries (ICIFI) comprised nine companies initially: Cow & Gate, Dumex, Meiji, Morinaga, Nestlé, Snow Brand, Wakado, Wyeth, and an unnamed ninth company.

The ninth company was Abbott/Ross, and although it had been involved in the discussions leading to the formation of ICIFI and was 'supportive of the concept', Abbott had some doubts about a proposed code of ethics that ICIFI was planning to publish. The ICIFI code did not specifically rule out the use of mass-media advertising for infant formula, a practice which Abbott said was 'inappropriate in third world settings'.[3] Abbott, therefore, stayed out of ICIFI.

The ICIFI code was made public two days after the end of the first hearing in the Nestlé trial. The timing of this, and of the formation of ICIFI, led War on Want to comment that the code was 'little more than a public relations exercise designed to stave off critics and deep public concern over their marketing methods'.[4] War on Want noted that, given the past record of the infant food manufacturers, what was required was not a weak, voluntary code, but 'a legal obligation on the companies, enforced by government and monitored by UN agencies'. War on Want was not alone in its criticism. The PAG in January 1976 said of the ICIFI code, 'The pledges do not go far enough and, as they stand, now pose a danger in that they provide for interpretations of the spirit of the code'.[5]

Criticism notwithstanding, it was a shrewd move on the part of the industry. It appeared as a positive and constructive step. Few people bothered to look at the ICIFI code of ethics, analyse the timing of ICIFI's formation, or the contradictions between the Council's objectives and the imperatives of commercial trading. For many observers, the message which came across was that the industry had voluntarily adopted controls on its marketing practices and had an international authority to police those controls.

Even the prestigious medical journal, *The Lancet*, which had uttered its share of criticism over the industry's practices, cautiously welcomed the formation of ICIFI.[6] It noted, however, that words alone would not suffice. 'It remains to emerge how far these principles, which the industry is more vigorously espousing, are convertible into actions which can surmount the dilemma of conflict between profit making and public health.'

Yet the principles themselves were sadly lacking in content. The ICIFI code contained *no* restrictions on existing promotion strategies. It did not rule out mass-media advertising, other public promotion, distribution of free samples, contact with mothers, or intensive promotion to health workers.[7] It merely legitimized those practices by permitting them to continue provided that all advertising and promotion *mentioned* that breast feeding was the 'first choice for the nutrition of infants'.

The penalty for failing to abide by the code was not very effective either. Adherence to the code was to be obligatory on member companies of ICIFI unless precluded by national laws. The penalty, should a member company 'persist in a clear violation of the code', would be the expulsion of the company from ICIFI.[8] Such a penalty would have no impact on the ability of the company to maintain sales or, indeed, to continue promotional practices that were incorrect.

The formation of ICIFI and the publication of the code was an attempt by the food industry to gain legitimacy. The industry could approach governments, health workers and the international agencies with the appearance of having recognized that there was a problem and that steps were being taken to resolve it. It was a strategy to gain control of the debate and move it out of the public arena, back into more genteel, more private discussions. Confirmation of this comes in ICIFI's first annual report:

> Special attention was given to the problem with what governmental organisations ICIFI should try to enter into official relations. There is no doubt that ICIFI would like to obtain a NGO status with the World Health Organisation. Furthermore, official relations with FAO would surely be of great value.[9]

Another lawsuit

ICIFI might well have succeeded in defusing the entire controversy had it not been for events that occurred in the United States involving Bristol–Myers, a non-ICIFI company. During 1974, the National Council of Churches (NCC) in the United States, concerned over the policies of a wide range of TNCs, established the Interfaith Centre on Corporate Responsibility (ICCR). The purpose of ICCR was to monitor, in particular, those TNCs in which the many churches and religious organizations affiliated to the NCC had investments. If any of the TNCs were found to be engaged in practices which went against the interests of society, shareholder power would be used in an attempt to modify the practice.

Given the impact of the 1974 World Food Conference, issues about food were considered most important by ICCR's board of management. Leah Margulies, the staff person assigned to ICCR's food task force had been involved in doing some research for Robert Ledogar's book, *Hungry for Profits*, particularly on the spread of bottle feeding in Latin America. That research had uncovered the involvement of American companies in the issue and it seemed to provide a good starting-point for challenging the activities of the companies. At the end of 1974, ICCR helped prepare shareholder resolutions on the major American infant formula companies. The resolutions were seeking, primarily, the disclosure of information about sales volume and marketing strategies in developing countries. Although none of the resolutions achieved a large enough vote at the annual general meetings to ensure that companies would comply with the demands, they were successful in raising the issues and began slowly to focus attention and pressure on the American companies.

With Bristol–Myers, however, the shareholder resolution had an interesting effect. Following the company's annual meeting in April 1975, a series of meetings were held between the company's management and the main proposer of the resolution, a Catholic order of nuns, the Sisters of the Precious Blood. Finally, in August 1975, Bristol–Myers issued a report on its marketing practices in countries outside the United States. However, the report failed to give any of the factual information requested. Instead, it was 'inaccurate, misleading, and apparently designed to confuse and mislead its readers'.[10]

The Sisters submitted a new shareholder resolution for the 1976 annual meeting requesting Bristol–Myers to provide, to all shareholders, sales and earnings data relating to infant formula by country and by product, and to provide information on promotional practices. The company issued a proxy statement urging shareholders to vote against the resolution.[11] In the proxy statement, Bristol–Myers claimed that its infant formula products

> are neither intended, nor promoted for private purchase where chronic poverty or ignorance could lead to product misuse or harmful effects . . . [They] are marketed through professional medical personnel and not directly to the consumer.[12]

Between the publication of Bristol–Myers' statement in August 1975 and the annual meeting in April 1976, several investors, including the Ford and Rockefeller Foundations, expressed disquiet with it and with the company's activities. A meeting between some of the investors and the company was held in December 1975. Bristol–Myers admitted that the statement had been compiled by its public relations department. Company representatives were unable to counter evidence presented which showed up factual errors in the statement.[13]

Despite repeated attempts by investors to enter into dialogue with the company prior to the April 1976 annual meeting to resolve the differences, the company persisted in its statements. The Sisters of the Precious Blood then filed a lawsuit in April 1976, charging Bristol–Myers with making misstatements in its proxy statement. Bristol–Myers responded with a motion to the court to have the case dismissed on the grounds that 'the subject of abuse in infant formula marketing practices is a matter for which no remedy can be provided by the US federal courts'.[14]

In order to counter Bristol–Myers' efforts to have the case dismissed, the Sisters and ICCR launched a world-wide investigation of the company's marketing practices. They received sworn affidavits from people in eighteen developing countries which showed clearly that

> Bristol–Myers was simply not telling the truth when it claimed in its 1976 Proxy Statement that its infant formula products are neither intended nor promoted for (A) private purchase where (B) chronic poverty could lead to (C) product misuse or (D) harmful effects, and that the products were marketed (E) through professional personnel and (F) not directly to the consumer.[15]

The case was dismissed in May 1977 because, according to the judge, the Sisters had not suffered 'irreparable harm' from anything that was published in the proxy statement. He declined to rule on the accuracy of the proxy statement. The Sisters were not satisifed and appealed. They asked other shareholders to support the appeal by filing 'Friend of Court' affidavits. Several shareholders agreed. So too did the US Securities and Exchange Commission (SEC), the statutory body which governs shareholder transactions.[16] Under US law, proxy statements cannot be incorrect or misleading. The SEC became concerned when the judge appeared to be establishing a precedent that a plaintiff in this type of case had first to demonstrate some economic disadvantage before any ruling would be made on the accuracy of a proxy statement.

During the rest of 1977, the Sisters had several meetings with Bristol–Myers. The talks made little progress until January 1978, shortly after the SEC did file a Friend of Court affidavit. At that point Bristol–Myers decided to settle the appeal out of court. The company agreed to send a report to shareholders which included the evidence collected by the Sisters. The company also agreed to halt all direct consumer advertising and to withdraw the use of mothercraft nurses.

The Bristol–Myers case was a significant factor in escalating the campaign. The legal requirement for carefully-documented evidence to disprove Bristol–Myers' claims tremendously increased the amount of data available about company practices. It also required further strengthening of links amongst the NGOs and health workers in many different countries. At the same time, much of this new information began to circulate amongst church groups and other organizations. The issue was raised in study groups to illustrate the need for greater corporate responsibility.

But while it provided for better understanding of the issue, there was one vital ingredient lacking—participation by ordinary people. Much of the work in the Bristol–Myers case involved lawyers sifting through the evidence and legal arguments. There was little that people who heard about the issue could actually do, aside from write to the companies involved and register a protest.

The Nestlé boycott

This frustration at not being able to do something led to another development in the campaign towards the end of 1976. Nestlé describes it as 'a campaign of action that was to prove considerably harsher than the most libelous of tracts'.[17]

In Minneapolis, the Third World Institute at the Newman Centre, part of the University of Minnesota, had organized study groups and various talks in and around the city to examine the issue. Many of the talks were illustrated with a documentary film, *Bottle Babies*, made in Kenya in 1975 by an independent German film-maker Peter Krieg.[18] After seeing the film and beginning to understand the conditions in which infant milks were being used in much of the world, and the tragic effects of their use, people spontaneously wanted to do something. At meetings one or two people would independently come up with an answer—stop buying the products of infant milk companies, and, in particular, Nestlé.

The idea began to gel. From a handful of people in Minneapolis a coordinating group was set up to organize a city-wide protest in July 1977. The group became the Infant Formula Action Coalition (INFACT), and on 4 July it officially launched a boycott of Nestlé.[19]

Nestlé was chosen for two reasons: it had the largest market share, so if its marketing strategies changed, other companies were liable to follow; and because it was Swiss-based, there was no other effective way of exerting direct pressure—American-based shareholder resolutions would not work.

INFACT made one basic demand on Nestlé: immediately halt *all* promotion of infant formulas, including direct advertising to consumers, the distribution of free samples to health care facilities and parents, the use of company 'milk nurses', and promotion to health workers.

By November 1977, groups across the United States had joined the boycott. It was supported by church organizations, student groups, trade unions, women's organizations, health workers, celebrities and some prominent politicians. In early 1978, the boycott spread to Canada, and later that year groups in New Zealand and Australia took it up.

Nestlé responded with a barrage of public relations efforts. It established an Office of Corporate Responsibility at its American head office in White Plains, New York. Hill and Knowlton, one of the world's largest public relations firms, was hired to improve the company's image. One of the first efforts by Hill and Knowlton backfired. A twenty-four-page glossy publication extolling the virtues of Nestlé was sent to some 300,000 clergy in the United States urging them not to take part in the boycott. For many of the clergy, it was the first they had heard of the boycott. Suspecting a guilty conscience, the clergy contacted INFACT for the other side of the story. By the end of 1978, many more church groups had endorsed the boycott.

Nestlé was genuinely surprised by the boycott, not the least because none of its infant formula products was sold in the United States.[20] Once again, Nestlé fell back on its strategy of questioning the motive of the groups and individuals involved in the campaign. The Nestlé Office of Corporate Responsibility issued a two-page leaflet which criticized the *Bottle Babies* film, implied that the World Council of Churches was involved and was conspiring to overthrow capitalism, and that individual campaigners who had also spoken out about the inequalities of the world's economic system were in some way a threat to society.[21]

Included in the leaflet was the inaccurate statement that Mike Muller, author of War on Want's report, *The Baby Killer*, was the 'leader' of War on Want and that the report had been 'an anti-Nestlé attack'. When War on Want drew Nestlé's attention to the facts, the managing director, Arthur Furer, agreed that errors had been made.

> You are perfectly right. As I very often said, your publication 'The Baby Killer' was not intended by 'War on Want' or used by this organisation as a direct anti-Nestlé publication. On the contrary ... I shall send a copy of this letter to our friends in White Plains as there seems to be a confusion concerning the publication of 'War on Want'.[22]

The early attempts by Nestlé to counter the boycott were fraught with such problems. In testimony before a Senate hearing in the United States in 1978, Oswaldo Ballarin, president of Nestlé Brazil, said, 'The US Nestlé Co. has advised me that their research indicates that this is actually an indirect attack on the free world's economic system'.[23]

Ballarin was virtually laughed out of the hearing. Senator Edward Kennedy, chairman of the hearing, told him 'the boycott is a recognized tool in a free economic democratic society'. Senator J. H. Chaffe added, 'I do not understand your objection to a boycott against US Nestlé operations: what better way of getting at the Swiss parent company?'[24]

A boycott is indeed a legitimate tool. Its purpose is to

> organise the non-cooperation of the consumers with the producers, in order to obligate the latter either to meet the demands of the workers who are struggling for the recognition of their rights, or to withdraw a noxious or harmful product from the market, or to improve the quality of a defective product.[25]

The object of a boycott is to apply pressure. Cesar Chavez, leader of the farm-workers in California, explained that pressure, not conversion of the opponents to a different point of view, was what got results. 'If we succeeded, it was thanks to the pressure. It will take a long time before the owners become converted. It is very hard to convert people when, between them and us, money is involved.'[26]

An understanding of how effective that pressure could be was driven home quickly to Nestlé. Two British communications experts noted that the impact of the boycott left Nestlé 'reeling from disbelief at the swift reversal of the public's perception of their role as benefactors of the malnourished to nothing better than arms salesmen'.[27]

Nestlé's public statements dismissed the boycott as 'symbolic' and 'an inappropriate measure'.[28] In 1980, Pierre Liotard-Vogt, chairman of the parent company, told the annual general meeting that the subject of the boycott would best be treated by 'being ignored since that is what it merits'. None the less, he went on to claim that 'the practical effect of the boycott can be considered to be nil, i.e. our sales have not been affected in the slightest, even at the height of the campaign'.[29]

He claimed, incorrectly, that the boycott was declining and was limited to the United States. In 1978, groups in Canada, New Zealand and Australia took it up. In 1980, the boycott spread to the United Kingdom, organized by the Baby Milk Action Coalition. In 1981, it started in Sweden and West Germany. In 1982, it spread to France. During 1983, organizations in Norway and Finland took it up.

Business International has called the international Nestlé boycott the 'most devastating' attack on the infant formula industry, one that 'will stand as an example for other campaigns'.[30] *Der Spiegel*, the West German magazine, called it 'a gigantic movement which has cost Nestlé millions of dollars'.[31] In 1981, a Nestlé-commissioned market research survey in Toronto, Canada found that 30 per cent of the city's 2 million inhabitants were aware of the boycott, with fully 10 per cent of the population actively boycotting the company's products.[32]

The publicity impact of the boycott has been tremendous. The economic impact is harder to quantify, not least because sales figures for particular market areas are never released by Nestlé. Nevertheless, some indications are possible. The Toronto example would suggest a loss of something in the region of $2 million annually, even taking a low average of $10 worth of purchases per person. The actual loss is liable to be higher, given that in mid-1981 the Toronto City Council voted to endorse the boycott and urged the removal of Nestlé products from all civic premises.[33] With strong boycott campaigns in Winnipeg and Vancouver in Canada, Boston, New York, Washington, Chicago, Cleveland, Minneapolis and Denver in the United States, and burgeoning campaigns in the United Kingdom, Sweden and West Germany, the estimate made by INFACT in 1984 that Nestlé had spent over $100 million fighting the issue seems a very conservative figure of what the boycott has really cost the company.[34]

In Chicago, in 1980, Nestlé was estimated to lose $15,000 as a result of a campaign to persuade schools not to use Nestlé chocolate bars in fundraising schemes. Efforts aimed at Nestlé-owned Stouffer hotels and the Rusty Scupper restaurant chain in the United States during 1980 also had economic impact. Protest pickets were set up outside several restaurants with an average of $50 worth of business an hour being turned away. At the hotels, organizations cancelled seminars and conferences, resulting in lost business worth hundreds of thousands of dollars. In the United States, Canada and the United Kingdom, many universities have taken Nestlé products off their campuses in recent years.[35]

The reality of the impact of the boycott is best expressed by a Nestlé manager in Canada, who said, 'Every time a customer goes into a store and makes a conscious decision not to buy one of our products, it hurts'.[36] Another clear indication of the impact of the boycott is the amount of time devoted to it by Nestlé executives. The top management of the company has been heavily involved in even mundane decisions over the boycott. In each of the countries where the boycott operated, executives poured out a constant stream of letters and other materials to people writing in protest over Nestlé's activities. For three days in 1980, Ernest Saunders, a senior vice-president of the parent company, held talks in the United States with the fifteen-member staff of the Office of Corporate Responsibility, the president of Nestlé US, the vice-president of Stouffers, the head of Nestlé Canada's public relations office, Nestlé's chief legal counsel in the United States, and representatives from the company's new public relations firm, the Daniel Edelman Agency. The object of the meeting was to devise a strategy for 'containment of the awareness of the boycott campaign, without being ourselves responsible for escalating awareness levels'.[37]

In a frank memo to managing director Furer, Saunders said, upon his return to Switzerland, that the basic strategies seemed adequate, but 'the *implementation* of the strategies have lost their "sparkle"'. The basic strategy Nestlé wanted to use was 'third party rebuttals of the activists' case'—the use of people who could not easily be traced to the company—to defend Nestlé's position. 'The best opportunity' for this and one that 'must be fully exploited' was seen to be an article that appeared earlier that year in *Fortune* magazine—'The corporation haters', written by *Fortune*'s Washington editor, Herman Nickel.[38]

Nickel's article tore into the campaign, claiming that it was run on emotion, not fact. He called the campaign 'a replay of the New Left of the protesting Sixties' and developed the theory that it was being largely directed by a Marxist-influenced coalition of church groups, particularly the National Council of Churches. 'Marxists marching under the banner of Christ' was the term coined in the article, and the overall message was that the controversy over infant feeding was unfounded, unnecessary and ideologically motivated. It was a plea to leave the corporations alone to get on with business.

Nestlé was so happy with the article that it considered using Nickel to further good advantage. Saunders said, 'It is important that in future work, the rebuttal of the unfounded attacks *on Nestlé itself* be more directly addressed, and I have also spoken to Mr Nickel about this. He will try to visit us in Vevey'.[39] Whether or not the visit was ever made is unclear. However, later developments suggest that Nickel was not prepared to be used as a propagandist for the company.[40]

In the meantime, Nestlé set to work to maximize the opportunity provided by the *Fortune* article. The article had its beginnings in 1979 when Nickel was approached by the Ethics and Public Policy Centre (EPPC), a Washington-based conservative think-tank founded by Ernest Lefever, to do a detailed study of the infant feeding issue. The study was never completed, however, and the EPPC decided to reprint the *Fortune* article and distribute it. Nestlé was already a financial contributor to the EPPC and was willing to help further. Saunders said, 'We should review the *optimum* mailing list for the *Fortune* reprint and then decide how to finance the operation'.[41]

This was coupled with ideas to place other articles selectively and to develop a 'so-called "grass roots" approach' with company executives armed with audio-visual material giving talks to local community leaders. Saunders summed up the Nestlé strategy this way:

> It is clear that we have an urgent need to develop an effective counter propaganda operation, with a network of appropriate consultants in key centres, knowledgeable in the technicalities of infant nutrition in developing countries, and with the appropriate contacts to get articles placed.[42]

The strategy was short-lived. On 4 January 1981, the memo Saunders had written was leaked to the *Washington Post*.[43] It was subsequently picked up by newspapers around the world and Nestlé was left red-faced. The manipulation the company had attempted was exposed.

There was, of course, another possible strategy that Nestlé could have used to end the boycott: stop the marketing practices that triggered off the boycott.

Involving the politicians

The publicity that grew up around the boycott had another effect. Not only did people who were boycotting write to Nestlé to let the company know of their displeasure, but many people also wrote to their Congressional representatives and senators asking them to take a stand on the issue. One of the senators who became interested in the issue was Edward Kennedy. He was chairman of the Senate Sub-Committee on Health and Scientific Research which had previously held hearings on the pharmaceutical industry. As the volume of mail about the infant-feeding issue grew, Kennedy decided to hold a hearing in May 1978 on the advertising, marketing, promotion and use of infant formula in developing countries.

Health workers with experience in developing countries, a representative from the WHO, a marketing expert, representatives from NGOs and executives from four companies—Abbott/Ross, Bristol–Myers, Nestlé and Wyeth—testified and submitted evidence. The hearing elevated the campaign in the United States and stimulated another wave of publicity. Public pressure had helped bring the hearing about, and as a result of the hearing, some of which was televised, more people heard about the issue and more pressure was generated. It vastly increased the amount of information available on the issue as well. A 1,500-page report of the hearing was crammed with argument and counter argument.[44]

Although the company representatives faced tough questioning from Kennedy and the other members of the committee, the conclusions of the hearing were little more than that more talks were needed. Kennedy recognized that the manner of infant-formula marketing in developing countries was adding to the enormous problems of infant illness and that the end-usage of formula products was not seen by the companies as a 'prime factor' in their marketing considerations. This lack of responsibility concerned him, and in June 1978 he organized a meeting with representatives from the industry to pursue the question further. He concluded that unilateral action by the United States to control the behaviour of the American-based companies 'might not be appropriate'.[45] He felt that an international solution to an international problem was required. He asked Dr Halfdan Mahler, director-general of WHO to convene an international meeting on the subject. ICIFI quickly notified WHO of its willingness to participate in such a meeting. As Nestlé put it, such a meeting would be the 'proper forum for discussion of all questions related to these problems'.[46]

WHO and UNICEF agreed to hold a meeting, and picked October 1979 as the date. Before the meeting took place, the industry exploited the 'proper forum' message, avoiding wherever possible any public debate and calling for an end to the boycott and other public protest. The point being stressed was that there was no longer any need for public concern. The international agencies would be able to deal with the matter and an amicable solution would be found. For the industry it was another opportunity to regain control of the controversy by moving from the public arena to a more private place. Or, in Nestlé's words, '. . . the industry wished to shift the discussion on infant formula marketing back to the sphere of relevant government authorities, health professionals and industry experts'.[47]

As it happened, the October 1979 meeting was a much more public gathering than the industry had anticipated. It proved to be an important forum for the NGOs and truly launched the internationalization of the campaign.

Notes

1. US Senate, *Marketing and Promotion of Infant Formula in the Developing Countries: Report of the US Senate Health and Scientific Research Sub-Committee*, Washington, DC, US Government Printing Office, 1978, p. 125.
2. P. S. Venkatachalam, 'The need for a regional/national industry council and how it would function', reprinted in US Senate, op. cit., pp. 1078–84.
3. D. A. Benton *et. al.*, 'Abbott Laboratories statement on infant malnutrition, breast-feeding and infant formulas in developing nations', reprinted in US Senate, op. cit., pp. 248–81.
4. War on Want press statement, reported by BBC TV programme *Panorama*, 1 December 1975.
5. A. Chetley, *The Baby Killer Scandal*, London, War on Want, 1979, p. 110.
6. *The Lancet*, 'The infant food industry', editorial, 10 June 1978, p. 1240.
7. ICIFI, *Code of Ethics and Professional Standards for Advertising, Product Information and Advisory Services for Breast-milk Substitutes*, Zurich, 1975 (amended 1976).
8. ICIFI, *Constitution and Articles of Association*, Zurich, 1975, (amended 1977, 1978 and 1979), p. 10.
9. ICIFI, *Annual Report* (covering the period November 1975–September 1976), Zurich, 1976, p. 3.
10. J. H. Churchill *et. al.*, *Memorandum of Law in Opposition to Defendant's Motion for Summary Judgement*, US District Court, Southern District of New York, 27 September 1976, p. 4.
11. A proxy statement is required by US law to be sent by companies to all shareholders, advising them of the agenda of the annual general meeting and of any proposed resolutions, setting out the arguments and any counter arguments to the resolutions. In this way, shareholders unable to attend the meeting can decide how they would vote and assign their vote to another person (proxy).
12. Churchill, op. cit., p. 23.
13. Ibid., p. 18.
14. ICCR, 'The bottle baby lawsuit', CIC Brief in *The Corporate Examiner*, April 1977, p. 3A.
15. Churchill, op. cit., p. 23–4.
16. Chetley, op. cit., p. 112.
17. M. McComas, G. Fookes and G. Taucher, *The Dilemma of Third World Nutrition: Nestlé and the Role of Infant Formula*, Vevey, Nestlé, 1983, p. 9.
18. The film itself soon became the subject of controversy. It was a powerful documentary which showed the appalling conditions in which infant formulas were used in Kenya and the distressing consequences. It included recordings of Nestlé radio advertisements and several shots of Nestlé promotional material. Several doctors at Kenyatta National Hospital in Nairobi participated in the film, including Professor N. O. Bwibo, Dean of the Department of Paediatrics. The Kenya paediatricians had an opportunity to vet the final script and made some minor changes that did not affect the basic thrust of the film. In approving the script, one paediatrician commented, 'We are delighted that the film will be a good one'.

    Just before the final print of the film was made in late 1975, the film-maker, Peter Krieg, decided to incorporate some additional information relating to the Nestlé libel trial. This was added into the last six minutes of the film and two versions were printed—one with the trial segment and one without. The Kenyan paediatricians saw the latter version in 1976 and ordered a copy for teaching purposes. Nestlé, meanwhile, responded by hiring another German film-maker to produce a film in competition—also filmed in Kenya and also using the staff at Kenyatta National Hospital. This was not sufficient for the company, however, and in January 1978, Geoffrey Fookes, an executive from the parent company in Switzerland, visited Dr Bwibo in Kenya and returned to state that Dr Bwibo was 'angry' about the Krieg film and would be writing a letter of 'condemnation' about it. Bwibo did write a letter, which he copied to Nestlé in April 1978, claiming the film 'did not represent the right aspects'. He said the film was 'emotional, biased and exaggerated'. He also claimed that he had not heard from Peter Krieg since the time the film was made.

Nestlé circulated the letter widely. Krieg subsequently disclosed all the correspondence related to the film and expressed 'shock and sadness' at Dr Bwibo's letter. Krieg concluded in a letter to Bwibo in May 1978: 'I can only assume that someone tried to deliberately mislead you or pressurise you to make a statement that is now being used in a way that could hardly have been your intention. I am sad to say that you are being used to support a case which I know is not yours.'

It was an example of the way in which Nestlé attempted to discredit its critics rather than deal with the facts behind the criticism. As a senior lecturer in paediatrics in Kenya advised one of the American NGOs involved in the campaign: 'As regards Prof. Bwibo's letter, I consider it unimportant, and you should not fear its effects . . . Don't get caught arguing *opinions*. Let Prof. Bwibo's opinions be. You have available to you in the literature so much *factual* information as to make this letter irrelevant.' Nevertheless, much time and energy *was* diverted from the main thrust of the campaign to document the incident and counter Nestlé's arguments in detail.

The film, incidentally, won the silver medal at the International Science and Technology Film Contest in Tokyo in 1976.

Full documentation on the incident appears in US Senate, op. cit., pp. 608–83. A summary appears in M. Karanja, 'Kenyans "star" in bottle-baby film', the *Standard*, (Kenya), 23 June 1982, pp. 11–12.

19. Chetley, op. cit., pp. 113–14.
20. Statement by Oswaldo Ballarin, president of Nestlé Brazil, in US Senate, op. cit., p. 128.
21. Nestlé, 'Infant formula in developing countries: a perspective', reprinted in US Senate, op. cit., pp. 872–3.
22. A. Furer, personal correspondence, 8 August 1979.
23. US Senate, op. cit., p. 127.
24. Ibid., p. 128.
25. J. M. Muller, 'Boycott actions and political pressure', in J. Wurst (ed.), *Moral Pressure for Liberation*, Antwerp, Non-Violent Alternatives and Peace Action News, 1981, p. 14.
26. Ibid., p. 15.
27. B. Beaver and F. Silvester, 'The gall in mother's milk: the infant formula controversy and the WHO Marketing Code', *Journal of Advertising*, **1**, 1, January–March 1982, p. 4.
28. H. G. Ciocca, 'The infant formula controversy: a Nestlé view', *Journal of Contemporary Business*, 7, 4, 1978.
29. P. Liotard-Vogt, Address given at Nestlé Annual General Meeting, Lausanne, 22 May 1980, pp. 10–11.
30. *Business International*, 'Corporate critics build steam for anti-MNC campaigns throughout Third World', 17 October 1980, p. 331.
31. *New Internationalist*, Issue 110, April 1982, p. 26.
32. *INFACT Update*, 'Canada: 10% Toronto residents actively boycott Nestlé', February 1982, p. 2.
33. *Globe and Mail* (Toronto), 'City boycotts Nestlé over baby formulas', 19 June 1981.
34. P. J. Hilts, '6½-year boycott of Nestlé is ended as firm adopts baby-formula code', *Washington Post*, 27 January 1984.
35. From various issues of *INFACT Update* during 1980 and 1981.
36. *INFACT Update*, November 1980, p. 1.

37. E. W. Saunders, memo to A. Furer, reproduced with annotations as *Nestlégate: Secret Memo Reveals Corporate Cover-up*, Cambridge, Baby Milk Action Coalition, 1981.
38. H. Nickel, 'The corporation haters', *Fortune*, 16 June 1980.
39. Saunders, op. cit.
40. Nickel's *Fortune* article was a spin-off from a study commissioned by the Ethics and Public Policy Centre (EPPC) which was delayed due to Nickel being involved in a motoring accident. Nickel appeared to be unaware that Nestlé was funding the EPPC. When he did find out, he refused to complete the study. For full details, see *US Senate, Hearings Before the Committee on Foreign Relations on Nomination of Ernest W. Lefever, to be Assistant Secretary of State for Human Rights and Humanitarian Affairs*, Washington, DC, US Government Printing Office, 1981, pp. 433–6.
41. Saunders, op. cit.
42. Ibid.
43. M. Mintz, 'Infant formula maker battles boycotters by painting them red', *Washington Post*, 4 January 1981.
44. US Senate (1978), op. cit.
45. P. T. Scott, 'Innocents abroad: infant food technology at the law's frontier', *Virginia Journal of International Law*, 20, 3, 1980, pp. 641–2.
46. A. Furer, personal correspondence, 14 August 1979.
47. McComas *et al*., op. cit., p. 14.

5. A unique meeting

> Poor infant feeding practices and their consequences are one of the world's major problems and a serious obstacle to social and economic development. Being to a great extent a man-made problem it must be considered a reproach to our science and technology and our social and economic structures, and a blot on our so-called development achievements. It is not only a problem of the developing world: it occurs in many parts of the developed world as well.
>
> from the *Statement and Recommendations of the WHO/UNICEF meeting on Infant and Young Child Feeding*, October 1979.

'There is a problem of infant malnutrition in the Third World, no one questions that, but we do deny that our company has in the past or is currently contributing to that problem', said John Stafford, executive vice-president of American Home Products, the parent company of Wyeth International.[1] He made the statement in February 1980, at a US Congressional Hearing before the Sub-Committee on International Economic Policy and Trade, which met to review the results of the October 1979 WHO/UNICEF meeting on Infant and Young Child Feeding.

His comment surprised Congressman Howard Wolpe who asked: 'So it is your testimony then that there is just no substantive problem, that the concerns that have been raised have arisen out of overreaction and emotional reaction that has no basis in reality?' Stafford replied:

> I think my testimony is that the industry, and particularly Wyeth, is very concerned about our products and our activities not only in the infant formula area but with respect to other products, and we want to eliminate any risk of a problem if we can. In my travels to the Third World I spoke with, for example, as I mentioned, the heads of hospitals in Kenya. I have been in Colombia, the Philippines, and I have asked administrators in hospitals, to describe to me in detail, the practices which we are charged with and asked them whether that is a problem. The answer invariably said: that is not a problem. We do not object to the way your company conducts itself, we do not think that is contributing to malnutrition in this country.[2]

Other health workers had different opinions. A leading South African paediatrician told one researcher, 'Wyeth is totally unscrupulous. Their advertising was bad to begin with and it remains so'.[3] A nurse at the Mater Misericordiae Hospital in Nairobi, Kenya was equally forthcoming:

> The S-26 (Wyeth formula) people bring samples twice a week . . . I think that without these companies we would solve a lot of these problems. But they keep

> coming, calling up, storming in with their milks. If these companies were not bringing their milks, all mothers would breast feed in the hospital and then go home and do the same. But to bring them to the hospital and expect us to give them to mothers. It is wrong. I get so annoyed![4]

These divergent opinions characterized the starting-points for the WHO/UNICEF meeting, with the industry maintaining that it had little or nothing to do with causing infant malnutrition or infant illness, and with health workers and pressure groups maintaining that the industry's activities, at the very least, exacerbated the problems.

WHO and UNICEF were expected by the industry to act as the moderators or referees in the dispute and 'depoliticise the controversy'.[5] John Donlan, vice-president and general counsel of Nestlé Enterprises, the American subsidiary of Nestlé, noted that:

> It was expected that by bringing this matter before an appropriate international forum, a clearer definition of the responsibilities of those involved in the formulation of infant feeding policy—including governments, the health professions and the infant food industry—would be reached. The dangerous over-simplification of issues regarding infant feeding propagated by activist groups made the convening of such a meeting a matter of urgency.[6]

Yet only three weeks before the meeting started, concern was being expressed by the industry over the outcome. Ian Barter, president of ICIFI, told *The Lancet*, 'We shall not be forced into accepting policies which we believe to be unjustified'.[7]

The reasons for the shift from a position of welcome to one of hostility were contained in the background paper which WHO and UNICEF prepared for the meeting.[8] It was sent out to the meeting's participants in early September and pulled no punches about the need to limit the use of artificial infant-feeding products. It noted that there was an 'excessive and inappropriate use' of the products, that their promotion had been 'more extensive and pervasive' than the provision of information about the benefits of breast feeding, and that unless 'strong interventions' were introduced to promote and support breast feeding, 'even larger numbers of infants and young children will be placed at risk of infections, malnutrition and death'.[9]

It identified the use of infant formulas as a 'major public health as well as an individual or family problem'. The paper said that 'only if breast milk is no longer available during the early months are they [infant formulas] necessary. Infant formulas are not needed during weaning'.[10]

The infant-food companies faced a massive decline in sales if those statements were to go unchallenged. ICIFI responded with a hard-hitting attack on the background paper as 'unscientific' and prepared its own 'Supplementary Background Paper' which was made available to all the delegates at the meeting.[11] The ICIFI paper claimed the statements made by WHO and UNICEF were 'in direct contradiction' with the PAG Statement No. 23:

> ... which should be the starting point for discussions ... The need for infant formula is far greater than that implied by WHO/UNICEF. There is no satisfactory evidence that the average mother in the Third World is able to breast feed her baby *exclusively* for the first four to six months of life without long term detriment to her baby's health.[12]

This is an important tactic to note, and one that was to be used over and over again by the industry any time there was a significant advance in the campaign. The idea was to refer back to something that had been negotiated previously, to claim that the companies were abiding by the negotiated agreement, and to protest that no one had any right to introduce a new position without the participation of the industry in drafting the position. It was elevating the requirements for free trade above all other considerations, such as the health and welfare of the consumers of the products.

ICIFI failed to mention that, in any event, PAG Statement No. 23 had been superseded. When the PAG was disbanded in 1977, it was replaced by the UN Consultative Group on Maternal and Young Child Nutrition, which in May 1979 offered the following advice about the use of infant formulas by mothers in poverty situations: 'They are expensive and, as a result, over-diluted feeds tend to be prepared. Thus, they are inaccessible in sufficient quantities for the target group discussed here'.[13]

The Consultative Group advised that:

> ... in the small percentage of cases where lactation fails during early infancy, the sequence should be to:
> 1. make all efforts to re-establish lactation;
> 2. provide breast feeding by a relative or wet nurse;
> 3. feed another milk or milk product available at the lowest price, including yoghurt;
> 4. feed a cereal augmented by any available milk, however limited;
> 5. feed a cereal augmented by other locally available protein sources.[14]

The Consultative Group, composed of some of the world's leading experts on nutrition and paediatrics, came to those conclusions free from commercial pressure. They found no lack of 'satisfactory evidence' to call for limits on the use of infant formulas.

Another factor leading to the industry's apprehension about the outcome of the meeting was the composition of participants. For the first time, people representing *all* the interested parties were to be sitting round the same table as equal participants. Twenty-three governments were represented; fourteen UN and specialized agencies, plus WHO and UNICEF; nineteen NGOs including the five most outspoken critics of the industry—AgDW, ICCR, INFACT, IOCU and War on Want; twenty-six representatives from the industry covering eleven of the major infant-formula companies; and twenty-two experts—specialists in nutrition, paediatrics, sociology, public health and marketing.

Thierry Lemaresquier of the UN Non-Governmental Liaison Service (NGLS) called the meeting

> a major departure from the traditional relationship between the United Nations and non-governmental organizations. For the first time, the public was itself recognised through several constituency-based organizations from both industrialized and Third World countries.[15]

According to *Business International*:

> ... their inclusion in this meeting came as something of a shock to the industry, which had understood NGOs to be rather different organizations—namely, anything but activists ... As WHO and UNICEF saw it, however, the 'public following' of the other groups was enough to warrant their inclusion in the discussions ... Moreover, the international organizations' officials were not sympathetic to industry's plaint over inclusion of the activists in the consultations, in light of companies' own lobbying power.[16]

Despite that, however, WHO, in particular, was extremely nervous about the meeting. As Edward Baer of ICCR explained: 'WHO had no experience in organizing such a meeting'.[17] Most WHO international seminars and meetings in the past had been technical meetings, with a few experts on a particular subject. Dr Halfdan Mahler, Director-General of WHO, commented after the meeting was over,

> WHO a couple of years ago would have vetoed any such kind of conference we have had with the non-governmentals in, with the consumers in, with the industry in, and with the Member States in, in a representative sample, and letting them loose with equal opportunity to express their convictions about a highly controversial subject.[18]

Baer adds that WHO had been under tremendous pressure from the industry prior to the meeting.

> The arguments the industry used were that WHO was making a drastic mistake in holding this type of meeting. They said it would undermine the scientific integrity of WHO and that WHO would lose its credibility by becoming political.[19]

With the industry suspicious of the pressure groups and of WHO's intent, and with WHO nervous about the behaviour of both the industry and the pressure groups, it was not surprising that the more active critics of the industry also had their worries. After five years of tough grass-roots campaigning and mobilizing public support, there was a real danger that the whole issue could disappear under a veneer of platitudes that would allow the industry to claim it had the WHO/UNICEF stamp of approval. By moving the issue into the international arena, it also moved directly into the realm of global politics, and WHO was not renowned for its political clout. True, World Health Assemblies in 1974 and 1978 had passed strong resolutions on the issue, but little had occurred as a result.[20]

On the other hand, the meeting was a precedent for establishing the legitimacy of popular organizations, and not simply those involved in the infant-feeding issue. If it worked, it would demonstrate that such organizations had important constituencies, concerns and inputs to bring to international negotiations. If it failed, it would be a very long time before such an experiment would be repeated.

Most importantly, however, as Dr T. A. Lambo, Deputy Director-General of WHO, said in his opening remarks to the meeting, 'the unmet needs of hundreds of millions of infants, young children, mothers and families' throughout the world were at stake.[21]

Dynamics of the meeting

Both the NGOs and the industry were busy immediately prior to the start of the meeting. The day before it opened, the NGOs arranged an 'information day' for any delegates who wished to attend. Presentations were made by health workers and campaigners. It was a fairly low-key exercise, but one which none the less increased the nervousness of WHO and UNICEF. They asked the NGOs not to hold the information day, claiming it could threaten the participation of the industry in the official meeting. One official went so far as to suggest that if the information day went ahead, it would be responsible for the death of millions of infants.[22]

The nervousness disappeared on the first morning of the meeting as people read through the *International Herald Tribune*. ICIFI bought page five and filled it with an advertisement claiming that better infant nutrition was a 'shared responsibility'—shared amongst the industry, governments and health workers.[23] It appeared as if ICIFI was *leading* the campaign for better infant health!

Delegates to the meeting were not impressed. It was a dramatic illustration of the way the industry operated and the attitudes of many of the scientists and government representatives began to harden over this display of insensitivity. The evening before the meeting started, ICIFI's 'hospitality suite' at the Inter-Continental Hotel was in full swing, and continued throughout the entire week. Veteran journalists in Geneva reported an unprecedented amount of wining and dining of delegates by the companies.[24] That, too, did little to impress people.

The third incident which aroused suspicion over the intent of the industry was its deployment of forces during the meeting. The meeting was structured with five working groups dealing with various aspects of the problems facing women and children. The first group dealt with the encouragement and support of breast feeding. Virtually every company had waxed long and loud about its support for breast feeding. However, only two industry representatives sat in on this working group. Neither of them made any contribution to the discussion.[25]

The second group dealt with the promotion and support of appropriate and timely complementary feeding (weaning) practices with the use of local food resources. As this group touched on commercial interests, six industry representatives were present, most of them from companies which manufactured weaning foods.

The third group discussed ways of strengthening the education, training and information available on infant and young child feeding. In its advertisement, ICIFI made a commitment to offer its 'expertise to Governments in educational programmes which promote sound infant nutrition'. However, only four industry representatives could be spared to take part in that discussion.

The fourth group, perhaps the most important because, in the words of one NGO participant, it 'represented the position of women',[26] looked at ways to improve the health and social status of women in relation to infant and young child feeding. It was not important to the industry. No one attended.

Most attention was focused on the fifth group and dealt with the appropriate marketing and distribution of infant-formula and weaning foods. Fourteen industry representatives took part.

While the first four groups proceeded smoothly and quickly with their discussion, the marketing group became bogged down. By the end of the first day of discussion, no progress had been made. It was a day of jostling for positions. The industry argued that the scientific premises on which the background paper was based were suspect and that more research needed to be done before decisions could be taken on how to resolve the problem. Attempts to redefine the terms used in the background paper took up most of an afternoon, with the industry proposing new definitions. For the NGOs in the working group, it was frustrating and discouraging.

In a planning session later that evening, with representatives from about a dozen NGOs, a few scientists and some of the government delegates, the day's events were reviewed. The progress made in the other working groups was encouraging. It became apparent that it would be unrealistic to expect the marketing group to arrive at consensus, so people in the other groups were urged to introduce some of the key marketing issues for discussion. The NGOs in the fifth working group were given the task of containing the industry—making sure that nothing disastrous occurred, but not expecting any results from the group.

The industry had been working that evening as well. The following morning, ICIFI introduced a paper it had prepared which outlined ten 'points of agreement' and five further proposals for discussion. It asked that the paper be distributed to delegates and that it should form the agenda for the day's discussion. It was an attempt to take control of the meeting, and it nearly succeeded. The chairperson of the group, Dr U. Frey from Switzerland, ordered the paper to be distributed. Delegate after delegate protested. WHO and UNICEF staff reminded Frey that there was only one official paper for the meeting, the background document. After an hour, Frey gave up in exasperation and refused to chair the working group any longer, saying it could discuss what it liked.[27]

After a short break, Frey was persuaded to remain in the chair. The industry paper was thrown out, and the meeting moved on. By the afternoon, the industry was feeling isolated in the marketing group. Governments, health workers and the NGOs were disputing virtually every statement made by the industry. Strong proposals were put forward for controlling marketing practices. Ian Barter, president of ICIFI, called the situation 'an abandonment of free enterprise and the exclusion of industry'.[28] Despite this, the working group was unable to agree on firm recommendations. As one government delegate put it, 'there are points to which only industry objected'.[29] All that the group as a whole could agree was a short statement which summarized the discussion. A smaller drafting group with representatives from industry, governments, experts and NGOs was set up to prepare specific recommendations.

At a final plenary session, the reports and recommendations from all the working groups were approved by consensus, although Nestlé sees the consensus 'as something of a misnomer, given that strong differences over precisely what should be considered "appropriate" marketing practices still existed'.[30] The recommendations dealt with the need for improved information, better practices in maternity wards, improvements in maternity benefits, the use of locally available weaning foods, better training for health workers, the use of mass media to promote breast feeding, the increased representation of women in the solutions to the problems, improvement in the nutritional status of women, and the control of marketing practices to the public and to health workers.[31]

On the last point, the meeting recommended that there should be no sales promotion to the public of 'products to be used as breast milk substitutes or bottle fed supplements and feeding bottles'. In an explanatory note, promotion was taken to include 'mass media and other forms of advertising directly to the mother or general public, designed to increase sales of breast milk substitutes, to the detriment of breast feeding'.[32] The important phrases in the explanatory note were 'designed to increase sales' and 'to the detriment of breast feeding'. They provided a loophole that was to be exploited following the meeting.

Two other important recommendations related to marketing: 'facilities of the health care system should never be used for the promotion of artificial feeding' and 'promotion to health personnel should be restricted to factual and ethical information'.[33]

These recommendations echoed earlier calls for controls on marketing that had been voiced by health workers, international agencies and pressure groups. They confirmed the validity of five years' campaigning. However, a mechanism was needed to put controls into place. The feeling that emerged during the meeting was that an international code was required. As a starting-point, delegates from Algeria and the United States put forward a modest proposal that 'there shall be no marketing or availability of infant formula or weaning food without being consistent with the ethical guidelines to be recommended by this meeting'.[34] Added to this were some basic elements that a code might contain.

There was general consent to this. The main objection came from the industry who began to talk of a loss of national sovereignty if an international code were to be developed! Barter explained, 'We welcome codes of conduct. We want to see as many codes as possible. The idea of a generalised code is one that we do not find helpful. An all-embracing international code can be unhelpful'.[35] It is worth remembering that, at that point, there already existed what the industry claimed was an 'international, all-embracing' code: the ICIFI code of ethics.

By the time the meeting concluded, it was agreed that 'there should be an international code of marketing of infant formula and other products to be used as breast milk substitutes'.[36] WHO and UNICEF were requested to organize the process for its preparation in consultation with all concerned parties.

The proposal developed by Algeria and the United States also found its way, in an amended manner, into the final recommendations. This paragraph, had it ever been enforced, would have put a remarkable constraint around the industry:

> There should be no marketing or availability of infant formula or weaning foods in a country unless marketing practice is in accordance with the national code or legislation if these exist, or, in their absence, with the spirit of this meeting and the recommendations contained in this report or with any agreed international code.[37]

Approval of this came towards the end of the final plenary session of the meeting, at a time when industry representatives were engrossed in the preparation of a press statement. It would have passed unnoticed by the industry, had it not been for a representative from an American company who woke out of his reverie to proclaim to his colleagues: 'My God, they've just committed us to every recommendation in this report!' By that time, it was too late; the chairman had moved on to the next paragraph.[38]

At the end of the meeting, everyone hailed it as a success. On behalf of ICIFI, Barter said,

> The meeting fulfilled its function admirably, which was to remove the important questions of infant nutrition from public polemics into constructive discussion in an impartial setting so that solutions could be found. From that point of view, it must be counted a great success.[39]

As Lemaresquier notes, before the meeting fears had been expressed that it would disintegrate into a confrontation between the industry and the NGOs, leaving everyone else on the sidelines with little hope of any sort of international action.[40] Martin Forman, director of the Office of Nutrition at the US Agency for International Development (USAID) and a member of the American delegation at the meeting, called it a 'unique opportunity' for many different groups to discuss the issues involved.

> Claims made by any group could be immediately challenged by a host of independently invited scientific experts. Government representatives were able

to query industry and critics on their views and on the evidence to support them. It was difficult to get by on reputation—or on generalities. Everyone and every comment could be immediately challenged—and usually was. As a result, everyone present was able to come away from the meeting with sharpened views of where everyone else stood on the issues. Participants had to stand up and be counted—and judged.[41]

Judgements about the NGOs were favourable. Both Mahler, director-general of WHO and Henry Labouisse, executive director of UNICEF, expressed praise for the role the NGOs played in the meeting.[42] One of the members of the American delegation said she had been impressed by the NGOs. 'I found the consumer activists to be most eloquent and very well versed with the problem.'[43] Lemaresquier saw the meeting as providing a new level of legitimacy for NGOs. He called it a 'confirmation of the very real socio-political power' that such organizations can wield.[44]

However, the ability to wield that type of power depends on several factors. Lemaresquier identified ten: three structural, four methodological and three attitudinal.[45] The structural factors involve the ability to build a constituency or primary support base that provides the first level of legitimacy; the ability to work in partnership with groups from other cultures or regions; and thirdly, the ability to generate support amongst other organized sectors of society that are not initially involved in a particular issue or concern. In simple terms, all three factors encompass an ability to build a successful interlocking *network* of groups and individuals who share some common perspectives.

With regard to working methods, firstly, research is important, and one could add analysis to that factor. Basic data about the issue are vital, but also needed is the analysis which demonstrates the interconnectedness of events and activities in different parts of the world and over the passage of time. An understanding of the process by which certain events occurred in the baby-food campaign was essential for gaining a clear perspective on the issue.

However, all the research and analysis is useless, unless something is done with it. Thus a second important method of working is the dissemination of information to a wider audience. Thirdly, it is important to use that information as a tool to lobby or negotiate for change. Finally, there is a need to monitor the results—in the case of baby foods, the impact that the campaign had on changing the attitudes and practices of both companies and governments and health workers. That information is then fed back into the process as a stimulant for further action. Indeed, *action* is the key word which summarizes the impact of these four factors.

Attitudinal factors are more difficult to identify, given that they deal with intangibles, with theoretical positions and to a certain extent with ideology. Lemaresquier said that one factor concerns NGO attitudes to issues which they take up and attitudes to their responsibility in being part of the solution.[46] The latter concept is particularly important, for by accepting the responsibility one accepts implicitly Lemaresquier's message that: 'If structures are to be changed, and you want to work towards that change, you have to challenge your own values and examine how much you yourself are part of the problem'.[47]

This, of course, presupposes awareness of a second attitudinal factor—the need for structural change and an integrated approach to these issues. Otherwise, it becomes all too easy to dismiss a particular concern as a social problem that can best be dealt with in isolation from other issues, or, indeed, in individual countries, rather than on a global basis.

The third attitudinal factor, therefore, is the need to recognize that issues such as the baby-foods one are only understood within the context of the complex interactions that exist between industrialized and developing countries. Thus, in this case, poor infant-feeding practices are not seen as only a problem for the developing countries. In short, these attitudinal factors lead to an *international* approach to issues.

Lemaresquier's analysis is important not only for the infant-feeding issue, but also for many other campaigns and issues that require socio-political change. In an increasingly independent world, failure to analyse the chains of cause and effect that influence everyday life makes it very difficult either to understand why events happen in a particular way or what action can be taken to encourage a more favourable course of events. In the infant-feeding issue, the ultimate decisions on how to feed rest with individual parents; however, the influences affecting those decisions are rooted, often virtually invisibly, in the global political and economic system. Without a global approach, without solid analysis and without the commitment for action, efforts at change are liable to be transient and superficial.

The NGOs attending the WHO/UNICEF meeting who were most critical of the industry possessed most of the factors which Lemaresquier identified as necessary for success. No organization had the same mix, the same proportions or exactly the same approach to the issue. But the blend that occurred during the meeting was effective. For the first time, many of these groups had an opportunity to work together. The sense of working in isolation was broken down and a strong team spirit developed.

In a spontaneous decision, representatives from six of the NGOs—AgDW, ICCR, INFACT, IOCU, Oxfam and War on Want—made a commitment to maintain the links that had developed, both personal and professional, and established a loose international information exchange. Originally called the International Baby Food Action Coalition, within a few weeks the name was changed to the International Baby Food Action Network (IBFAN). It was following on from the attempt that had been made after the Berne meeting in 1976 to establish an information exchange, but IBFAN was to succeed where the other had failed, largely because of the experience of its founders having come through the WHO/UNICEF meeting. The meeting had been intense and stressful, and the NGOs found that they had to trust each other and work together. They needed each other to succeed.

Although none of the founders recognized it at the time, the birth of IBFAN was NGO history in the making. Admittedly, there were already many international NGO groupings in existence and several of the NGOs at the Geneva meeting were involved in them, but IBFAN was different. It could be best described as an 'illusion', in the sense that it had no fundamental organization, no officials, no formal membership.[48] It both existed and did not exist at the same time.

The first agreed objective of IBFAN was to act as a monitor of corporate activity around the world, using the recommendations of the October 1979 meeting as a guideline, and to share the information gathered as widely as possible. With no staff and no funds, the work of monitoring and the cost of information-sharing was to be borne by the founding groups, as part of their overall activity on the campaign. Basically, IBFAN survived its first year on the commitment of the people who founded it. It was not without growing pains.

A small coordinating meeting held in September 1980 noted that 'things are done on an *ad hoc* basis. Decision-making in IBFAN is now dependent on who is where at what time. This has led to some confusion at times.'[49] However, it was felt that the confusion was not all that serious.

> Basically, when it comes to IBFAN decisions which can affect our position on an international level, there are about 10 of us involved and we all trust each other. This is a fundamental asset which industry does not have ... From a tactical point of view, there is a definite advantage in being a loose network of several organizations from all over the world.[50]

For a confusing illusion, IBFAN had a powerful impact during its first year of life. It made press headlines. It published reports of the industry's violations of the October recommendations. It began producing a regular monthly newsletter. It stimulated the involvement of more groups and individuals in the issue. It developed good contacts with some government officials in different countries.

The issue was becoming truly international. Public concern had started in the United Kingdom with War on Want's report, then spread to Switzerland and Europe with the translation of the report by AgDW. Then it crossed the Atlantic to the United States with the work of ICCR on the American companies and INFACT with the Nestlé boycott. The October 1979 meeting brought the focus back to Europe and then it began to diffuse throughout the world. IBFAN helped that process by bringing together contacts in different countries. It stimulated a series of small activities that meant, for the first time, that TNCs had to fight on several different battlegrounds at once.

Groups in Africa, Asia, Latin America, the Pacific, North America and Europe were getting fresh information. They were all asking the same questions, making the same demands, and using a variety of ways to get the message across: promotion of artificial infant feeding must be stopped.

Far from containing the issue, the October 1979 meeting set off a series of brush fires all over the world, fuelled to some extent by the improved flow of information stimulated by the small number of groups who started IBFAN. It was a headache for the industry. In August 1980, Ernest Saunders, the new president of ICIFI and vice-president of Nestlé, explained how effective this approach was.

> In view of the overall propaganda campaign now being mounted through IBFAN and the professionalism of the forces involved, it is always possible that we could even win a battle in the US and lose the war as a result of the determined pressure on Third World governments and medical authorities.[51]

The main focus for that battle was to be the international code being developed by WHO and UNICEF. Saunders' prediction was to come true, although at the time neither he nor anyone else could have imagined the bizarre turn of events that would, once and for all, put the infant-feeding issue firmly and fully in the public eye.

Notes

1. US Congress, *Marketing and Promotion of Infant Formula in Developing Countries: Hearings Before the Sub-Committee on International Economic Policy and Trade of the Committee on Foreign Affairs of the House of Representatives*, Washington, DC, US Government Printing Office, 1980, p. 88.
2. Ibid., p. 89.
3. Ibid., p. 102.
4. Ibid., pp. 179–80.
5. M. McComas, G. Fookes and G. Taucher, *The Dilemma of Third World Nutrition: Nestlé and the Role of Infant Formula*, Vevey, Nestlé, 1983, p. 14.
6. US Congress, op. cit., p. 80.
7. *The Lancet*, 'Infant food industry's misgivings about next month's Geneva meeting', 22 September 1979, p. 649.
8. WHO/UNICEF, *Joint WHO/UNICEF meeting on Infant and Young Child Feeding, Background Paper and Themes for Discussion* (Doc. FHE/ICF/79.3), Geneva, 1979.
9. Ibid., pp. 32, 35.
10. Ibid., pp. 59, 43.
11. ICIFI, *Supplementary Background Paper for the Meeting on Infant and Young Child Feeding*, Zurich, 1979.
12. Ibid., p. 23. (The ICIFI claim that there was no satisfactory evidence that Third World women could adequately breast feed for the first four to six months was absolute nonsense. Millions of mothers all over the world were providing adequate nutrition for their infants through breast feeding. Even amongst undernourished mothers, the evidence existed to show that the volume and composition of human milk was 'surprisingly good' (Jelliffe and Jelliffe, *Human Milk in the Modern World*, p. 80). One particular study in Nigeria during drought and famine conditions in 1974 found that 'mothers, undernourished by non-famine standards of contiguous African nations were able to provide enough breast milk for normal growth of their infants to six months of age without detriment to their health' (J. Murray and A. Murray, 'Breast milk and weights of Nigerian mothers and their infants', *American Journal of Clinical Nutrition*, **32**, 4, April 1977, p. 737). A more recent study in Brazil confirms this: 'We conclude that low-income mothers' milk contains enough energy to sustain growth of their

infants at acceptable levels for the first six months, even when the women were malnourished and the babies exclusively breast fed' (P. C. Marin *et al.*, 'Energy content of breast milk of poor Brazilian mothers', *The Lancet*, 28 January 1984). In any event, the solution to the problem of poor maternal nutrition is not to introduce an expensive powdered milk supplement for the baby, but to concentrate on interventions that will ensure that the mother has access to adequate nutritious food.)

13. United Nations University, 'Consultative group on maternal and young child nutrition', *Food and Nutrition Bulletin*, **1**, 3, May 1979, p. 20.
14. Some insight into the lengths to which the infant food industry was prepared to go to play down the importance of the advice given by the Consultative Group came to light during the US Congressional Hearings in January and February 1980, when Edward Baer of ICCR referred to the list of steps to be taken in the event of lactation failure. Thomas Ward, Nestlé's legal counsel in the United States, subsequently submitted a letter to the chairman of the Hearings stating: 'We should like to caution the sub-committee about taking Mr Baer's interpretation of the report by the UN consultative group on maternal and child feeding . . . at face value. E.g., in answer to a question from the chairman . . ., Mr Baer made the extraordinary statement that the group is composed "mainly of paediatricians from Sri Lanka"'.

    What Baer actually said in response to the question was, 'Mainly paediatricians. The chairperson is Dr Piryani Soysa from Sri Lanka and the committee includes about a dozen representatives from different countries . . .'. (For full details, see US Congress, op. cit., pp. 97–8, 162.)
15. T. Lemaresquier, 'Beyond infant feeding: the case for another relationship between NGOs and the United Nations system', *Development Dialogue*, **1**, 1980, p. 120.
16. M. McComas, *Europe's Consumer Movement: Key Issues and Corporate Responses*, Geneva, Business International, 1980, p. I–84.
17. E. C. Baer, personal communication.
18. Statement by Halfdan Mahler at press briefing, Geneva, 12 October 1979.
19. Baer, op. cit.
20. Resolution WHA27.43 and Resolution WHA31.47 in *Handbook of Resolutions and Decisions of the World Health Assembly and Executive Board*, Volume II, 4th edn., Geneva, WHO, 1981, pp. 58, 62.
21. A. Chetley, *The Baby Killer Scandal*, London, War on Want, 1979, p. 161.
22. Personal communication from ICCR.
23. ICIFI, advertisement in *International Herald Tribune*, 9 October 1979, p. 5.
24. US Congress, op. cit., p. 53.
25. Chetley, op. cit., p. 163.
26. US Congress, op. cit., p. 68.
27. Chetley, op. cit., pp. 165–6.
28. Ibid., p. 168.
29. Ibid.
30. McComas *et al.*, op. cit., p. 14.
31. WHO, *Joint WHO/UNICEF meeting on Infant and Young Child Feeding: Statement and Recommendations*, Geneva, 1979.
32. Ibid., p. 28.
33. Ibid., p. 29.
34. Chetley, op. cit., p. 167.
35. Ibid., p. 168.
36. WHO, *Statement and Recommendations*, op. cit., p. 29.
37. Ibid.
38. Chetley, op. cit., p. 170.

39. US Congress, op. cit., p. 82.
40. Lemaresquier, op. cit., p. 122.
41. US Congress, op. cit., p. 10.
42. Chetley, op. cit., p. 171. See also A. Chetley, 'The baby milk controversy: a rejoinder', *Food Policy*, August 1980, p. 227.
43. US Congress, op. cit., p. 13.
44. Lemaresquier, op. cit., p. 121.
45. Ibid.
46. Ibid., p. 123.
47. Ibid., p. 124.
48. IBFAN, *Report of the First Congress*, Geneva and Minneapolis, 1981.
49. IBFAN, *Report of a meeting of IBFAN* (mimeo), Geneva, 28 September 1980, p. 5.
50. Ibid.
51. Baby Milk Action Coalition, *Nestlégate: Secret Memo Reveals Corporate Cover-up*, Cambridge (UK), 1981.

6. Making a code

> The longer these codes are discussed, the more moderate and practical they seem to become.
>
> Arthur Furer, president of Nestlé[1]

Euphoria reigned at the end of the October 1979 meeting. There was no doubt that a significant step towards the control of the marketing of infant-feeding products had been achieved. While admitting that there was still more to be done, WHO director-general Mahler told a press conference, perhaps naïvely:

> If you read these recommendations I find it very difficult to see how industry could leave this meeting with any ambiguities ... I don't think in any way industry could get away with that and say now we have got WHO approval for doing what we have done in the past.

Within a week of the meeting that was precisely what the industry *was* saying. Arthur Furer of Nestlé told the Swiss newspaper, *Tages Anzeiger*:

> We do not feel restricted in any way in the field of commercial activity by the WHO recommendations. On the contrary. The gradual changes that Nestlé has introduced in the past five years are completely in line with these recommendations.[2]

In the months immediately following the October meeting, the infant-food industry publicized its own loose interpretations of the recommendations. By not accepting the fundamental premise upon which the recommendations were based—that marketing practices are a contributory factor in the decline of breast feeding and thus poorer infant health—the industry was able to sidestep the spirit of the meeting and its recommendations and warp the wording to suit its own interests.

Thus Furer was able to claim that advertising was permitted by the recommendations, provided it was of 'an educational nature'.[3] David Cox of Abbott/Ross was able to claim that the recommendations did not rule against the giving of free samples, provided they were given to health workers to pass on, rather than directly to mothers.[4] John Stafford from American Home Products/Wyeth interpreted the recommendations to mean that the company could continue to make available materials for mothers such as its 'Baby Book', despite those materials containing advertising and promotion for formulas.[5] As late as May 1980, Nestlé maintained that it did not believe that any additional changes in its policy were called for as a result of the WHO/UNICEF recommendations.[6]

By sticking to the most limited interpretation of the recommendations, the industry was attempting, once again, to contain the process aimed at controlling unnecessary marketing practices. It was a way of drawing a line beyond which the companies were not prepared to go. Cox explained that:

> ... when we came away from Geneva, within the interpretations that we felt were the consensus of that meeting, we felt that we could live with the recommendations ... However, in conversations with the staff of WHO and UNICEF since the Geneva meeting, we have had the individual interpretations of how they could interpret those recommendations, and some of the interpretations we cannot live with.[7]

The industry refused to move forward without protracted negotiation. Each attempt to move the debate along was resisted strongly, with the industry falling back to previously-agreed statements. The NGOs, on the other hand, were constantly urging the process on. They were the catalysts for change and saw each agreed position as merely another step in a long process.

With the NGOs and the industry pulling in opposite directions, WHO and UNICEF adopted the role of mediators. Officials at both organizations saw themselves and their agencies as the voices of reason—a conciliation service that could maintain a neutral stance.

It was a position that infuriated and frustrated the NGOs, who had assumed that the international agencies, particularly WHO, were combatants in the struggle for better infant health, not the referees. The NGOs saw their own role as supporting the work of WHO. The idea that WHO should act as 'honest brokers' in a negotiated code was unacceptable to the NGOs. Leah Margulies of ICCR explained the position of the NGOs:

> We feel the code should clearly state what are the best kinds of practices to ensure infant health and it should not be arrived at through a negotiation where the companies say, all right, we will stop such and such a practice if you will take out of the code another kind of restriction. We feel that it is the job of the international health agencies to articulate the finest possible code of conduct to protect infant lives.[8]

The first draft of the code, prepared by WHO and UNICEF staff and published in February 1980, attempted to set down an ideal series of provisions. It included many of the suggestions that had been proposed in the background paper for the October meeting as well as those proposed during the meeting. WHO staff felt that the code was close to its final version and that a series of consultations with all the interested parties during February and March 1980 would resolve any anomalies so that the code could be presented as a final document to the May 1980 World Health Assembly (WHA) for ratification.

The three American companies, Abbott/Ross, Bristol–Myers and Wyeth, found the first draft far from ideal. They said it was 'seriously defective by almost any measure. It contained a long series of unwarranted and absolute prohibitions and restrictions of legitimate commercial activity'.[9] An ICIFI spokesman called the code 'wholly irresponsible' and claimed that if it was approved, 'it would certainly kill hundreds and hundreds of babies'.[10]

A consultation with experts in early February 1980 led to a reaction that the code was a good start, but required more specificity and precision in order to be effective. The experts also felt that it would be unrealistic to attempt to have the code ratified at the WHA in May, and suggested the following year's WHA as the target date.[11]

A week later, NGOs expressed similar views at a consultation. They felt that the code had strengths but noted there were contradictions between some sections; that there was a lack of clear definition in places; and that controls on some particular practices had been omitted.[12]

Nearly fifty governments sent responses on the code to WHO. The majority were in favour of the draft code and, not surprisngly, these were almost all developing countries. The exceptions were Belgium, which called the code 'highly justified' and offered a few suggestions for making it stronger; Bulgaria, which called it 'a timely and advisable initiative'; Finland, 'acceptable in general'; Greece, 'in total agreement'; and the Soviet Union, 'desirable and timely'. In contrast, the major milk-exporting countries—Denmark, France, West Germany, Netherlands, New Zealand, Switzerland and the United States—rejected the first draft.[13]

The American position was undoubtedly helped by some sharp lobbying organized by Wyeth. The company hired Bill Rogers, a former under-secretary of state to arrange a private briefing with officials in the State Department. The briefing was held on 20 March.[14] Nine days later, without consulting any of the pressure groups, breast-feeding promotion groups or health workers from the United States who had attended the October meeting, the United States' government sent a letter to WHO calling for a code of general nature which was to be voluntary or recommendatory. 'A document with provisions that are mandatory on governments or corporations could not be accepted by the United States,' said the letter. The United States also warned that:

> The decision to proceed with a "code" or formal standards should be a deliberate step of the Assembly, one that can provide guidance for the approach to be taken. The contents of the document must then be subject to full intergovernmental negotiations.[15]

The proposal for an intergovernmental group to negotiate a code was a direct slap in the face to WHO and UNICEF, suggesting that they had overstepped their mandates. At the same time, for anyone who was opposed to a tough code the idea of intergovernmental negotiations was attractive. It would mean a much slower process: the UN had been negotiating a code of conduct for TNCs for over five years, and major stumbling-blocks still remained in it. In addition, as negotiations drag on, the provisions of any such code are more likely to become more generalized, weaker and ineffective. From the perspective of a TNC this would be helpful, for while negotiations were in progress, TNCs could argue that it would be improper to implement any of the proposed provisions until *all* the provisions had been agreed.

A second draft of the code was prepared in time for the WHA in May 1980. It received little detailed discussion, however, as the main efforts were concentrated on gaining acceptance for the October recommendations and mandating WHO to continue with the code development process.

WHO officials were upset over the United States' government's stance, not least because the United States contributes 25 per cent of WHO's regular budget. They felt, however, that a resolution being prepared by the Swedish delegation with the support of other Scandinavian countries and some prominent developing countries would deal with the problem. During the first week of the three-week assembly, support for that resolution continued to grow, although there were fears that the United States might put forward a counter proposal. By the end of that week, the United States decided to join in support for the draft resolution, knowing that before it was finally approved, there would be a small drafting group established to incorporate amendments.

Discussion on the issue came up at the end of the second week with fifty-four countries speaking. Early on in the discussion, Dr Mahler explained to the delegates that he had been accused of making WHO play a role it should not be playing and had overstepped his mandate as WHO's director-general. He referred the delegates to the resolutions passed by the WHA in 1974 and 1978 and said his conscience was clear that he had not exceeded his mandate. 'Obviously, whenever raw political or economic nerves of society are touched, explosions result. I am not to be envied for the letters I have received. Nevertheless, it is part of my job to expect such attacks.'[16]

His intervention helped to ensure favourable comments from everyone who spoke. Burma, for example, expressed gratitude for the 'steps taken by WHO and UNICEF to prevent the exploitation by multinational companies of poor countries like Burma . . . We welcome the move to establish an international code of breast milk substitutes and regret that WHO had not taken action 25 years ago'.[17] Similarly, the delegate from India said:

> I hope no effort will be made to dilute the WHO/UNICEF recommendations because of subtle pressures by multinational and even national infant food industries who have already done enough harm to many generations of children, particularly in developing countries.[18]

As was expected, a drafting group, which met in closed session, was established to finalize the resolution to adopt the recommendations of the October meeting. Ten countries were chosen to sit on the drafting group, representing a good balance between developing and industrialized countries.[19] However, like all such drafting groups at WHAs, it was left open-ended—other countries could sit in if they wanted. Among the countries who chose to do so were the major milk-producing countries: Australia, Denmark, West Germany, New Zealand, Switzerland and the United Kingdom. With such a heavy weighting in favour of the industrialized countries, the drafting group proceeded very slowly. According to members of the group, the United States played a very low-key role. The United Kingdom, New Zealand and France were reported as the main instigators of passages that would ensure that the possibility of a binding code was ruled out. A last-minute attempt by the United States to introduce the idea of intergovernmental negotiation on the code was expected and the Scandinavian delegates were prepared to squash such a move within the group. It never came. Finally after three days, consensus was reached and the resolution was finalized for presentation to the assembly on a Thursday afternoon, less than twenty-four hours before the WHA was due to end.[20]

The asembly had originally been scheduled to conclude on the Thursday, but had been extended an extra day. However, many delegations had either left already or were preparing to do so. With only a few formal resolutions to vote on and then the closing ceremonies, most delegations were in a very relaxed mood.

Half an hour before the afternoon session was due to start, the news broke that, despite having participated in the drafting consensus on the resolution, the United States intended to move two amendments to the resolution which would insert the intergovernmental negotiations idea. Many delegates who sat through long hours on the drafting group for the resolution were furious at this last-minute attempt to upset the process. Dr J. Bryant, who introduced the amendment for the United States, told the delegates:

> It is with the greatest reluctance that my delegation introduces these amendments. We apologize to our colleagues in the drafting group, but I have just now been instructed by my government to express concern over the mechanisms for the development of the code. My government does not believe the resolution is explicit enough on this point. When governments are asked to agree a code, governments should formulate the code.[21]

Rather than allowing for discussion to open up on the amendments—something which could have meant delaying the whole process for a year—the Norwegian delegation called for an immediate vote. The amendments were defeated and WHO received the necessary approval to carry on with the code. But a final warning note was sounded. Following the vote on the resolution, Bryant announced:

> I would like the record to show that although my government has joined the consensus on this resolution, we express serious concern with respect to the mechanism. My government does not want this decision to be considered as a precedent for the drafting of codes in other UN agencies. The US delegation believes that agreements of this nature between governments should be

> negotiated between governments. That is the practice regarding preparation of codes in ECOSOC, UNCTAD, the ILO and other parts of the UN system.[22]

Canada placed similar reservations on record.

The possible explosion Mahler had mentioned at the beginning of the discussion did not quite materialize. In its place was a quiet tension—a sense that an even bigger storm was yet to come.

It was now clear that WHO and UNICEF would continue the code development process. It was clear that a code would emerge at the end of that process. The contents of such a code were up for discussion, but it was unlikely that there would be much in the code that would give comfort to the infant-food industry or to the industrialized milk-exporting countries. While detailed discussion continued on the code, the major question that began to emerge was, what legitimacy would the code have?

WHO had three possible methods for bringing the code into force, according to its constitution:

— a convention, which requires a two-thirds majority of the WHA to become adopted, and which comes into force when it is ratified by each member state (member states have eighteen months to decide whether or not to ratify);
— a regulation, which requires a simple majority of the WHA and comes into force when accepted by member states within a stated time period, usually no more than twelve months (as with the convention, member states may opt out, providing they notify WHO in writing of the reasons for doing so);
— a recommendation, approved by either a simple majority or consensus agreement of the WHA. It is then up to member states to act or not act upon such recommendations. There is no specified time limit and no necessity to act.[23]

The secretariat of WHO had no intention of trying for a convention. It would take too long to come into force, and would involve too much political manoeuvring to ensure getting the necessary two-thirds vote at the WHA. The choice basically was between regulation and recommendation. During the early part of 1980, there was a strong lobby *within* the staff of WHO and UNICEF in favour of a regulation. They saw the issue as too important to be left up in the air and wanted to see definite action.

During the 1980 WHA, there was little discussion on this point. Only eight countries mentioned it directly—four in favour of regulation, four for recommendation—and another dozen hinted broadly, again with equal weight on either side. Following the WHA, written comments on the second draft of the code were sent to WHO by governments, the industry, the NGOs, UN agencies and experts. A third draft was prepared in time for two more consultations in August and September. The first consultation was a mini-version of the October 1979 meeting with the exception that no governments were present. They attended the second consultation.

The industry's positions on the second and third drafts were not that different from their opinion of the first draft. According to the three American companies:

> Although the second draft was described by the WHO as 'a series of principles', it continued to contain the same kind of unwarranted, unrealistic and otherwise irrelevant and arbitrary restrictions, prohibitions and requirements which characterized the first draft.
>
> In all substantial respects, it (the third draft) continued to suffer from the same types of problems which characterized the first and second drafts. A careful examination of the third draft would reveal an instrument still designed more to impose arbitrary and needlessly severe restrictions on commercial activity than to promote sound infant nutrition.[24]

The industry prefaced the August consultation with a long meeting with WHO and UNICEF staff at which it made a series of ten demands about the code—mainly complaining that it was too negative—protested over the inclusion of NGOs in the consultation, and threatened to walk out if the NGOs became confrontational.[25] WHO and UNICEF listened patiently (as they listened patiently to NGOs who had a meeting with them after the industry). They pointed out to both groups that the consultation was not a negotiating forum. They expected differences of opinion and were not attempting to resolve those differences, but merely record them, for consideration in preparation of the fourth draft.

WHO was smarting somewhat over the inclusion on ICIFI's team at the consultation of the industry's newest acquisition, Dr Stanislas Flache, a former assistant director-general of WHO. Flache retired from WHO on 31 July 1980. The next day, he began his new job as secretary general of ICIFI. *Business International* said of the acquisition:

> That choice was obviously not haphazard. Suspecting that international organisations such as WHO and UNICEF may harbour their own brand of activists, industry can afford to be nothing less than meticulous in the way it deals with these bodies. An inside track will probably help.[26]

The consultation was not a smooth one for the industry. It found almost no support for its positions and by the second day of the consultation was simply saying 'we will submit in writing our views on this issue'. An internal War on Want memo on the consultation said:

> The main industry delegation helped to polarise the interpretations of the code with amazing clarity. It basically led to an agreement to disagree over everything from the title of the code to where the full stops should come. For anyone who had been involved in the code process since its inception, it was a tedious meeting. However, for the new faces, and there were some among the scientists and UN agencies, it was an eye-opening education exercise. Many of the

> previously non-committed experts, UN agencies and even the milder NGOs were taking stronger stands on the basic principles of the code, a process which will help to strengthen the hand of WHO/UNICEF in drafting the code.

There was no discussion on the regulation/recommendation question as WHO and UNICEF noted that this would be decided by *governments* at the WHA in May 1981. It was, therefore, somewhat surprising that the September consultation amongst governments failed even to raise the question. The reason for that is explained in the cable sent from the American mission in Geneva back to Washington at the end of the meeting (many official cables were leaked to INFACT). It said, 'the issue of most importance to USA, whether the code is an Article 21 "regulation" or an Article 23 "recommendation", was kept off the formal agenda by tacit agreement of the participants'. It was, however, widely discussed elsewhere, as the cable reveals:

> The question of the code's legal nature was the subject of extensive discussion, initiated by USdels, in the corridors. USdels sought views of great many countries and learned that a number of LDCs, including Lesotho, Jamaica, India and the Philippines favoured recommendations. USdels indicated unmistakably in such conversations the importance placed by the US on proceeding by way of a recommendation.

The American delegation had a very busy time at the consultation. It met with an IBFAN team that was present in a lobbying capacity; met several times with WHO and UNICEF staff; met with the industrialized countries present at the consultation; and, finally, had a luncheon for all the delegations.

Edward Baer of ICCR was part of the IBFAN team which met with the American delegation. He said that Dr Bryant, head of the delegation, explained that the United States would accept certain controversial passages in the code, providing it was going to be a recommendation. However, if it was to be a regulation, Bryant said it would change the position of the United States' government entirely. Baer commented:

> So I asked Bryant, 'I see, there are two separate, completely different sets of standards for how best to protect babies' lives? And if the code is going to have more binding authority as a regulation, then the US thinks there is an entirely different set of standards that would have to apply?' Bryant was furious, and said, 'No, that's no true. You're putting words in my mouth.' But basically, what was being suggested here was that the US was only willing to say fairly progressive things if they knew that the legal basis of the code would give them all the room for manoeuvring and for interpretation as to render the code less meaningful or irrelevant.[27]

Baer was alarmed as well by the information he found out about discussions amongst the industrialized countries on the afternoon before the official consultation began.

> Their basic position was this: anytime this question would come to a vote at a UN forum like the WHA, the industrialized countries would lose because there are simply more Third World votes than there are industrialized country votes and therefore, the afternoon on Wednesday was spent discussing how they could outfox the Third World—to find a way of not having to abide by any of the recommendations in a legally binding form. The whole sense of that meeting was 'we don't give a damn about what happens over there. What we give a damn about is the trade rights and the commercial rights of our companies to sell, promote and market their products in any way that they see fit. There was a complete and blatant lack of respect for whatever health, nutritional and economic problems the promotion might have on people in the Third World.[28]

Much of Baer's analysis was proved right from the contents of the cable sent back to Washington by the American mission in Geneva after the consultation. Commenting on the meeting with the industrialized countries, the cable noted that the legal nature of the code was 'the most important issue for almost all the countries represented'. The cable said that only Sweden had a strong position in favour of a regulation. The rest were in favour of recommendation. The meeting involved Australia, Canada, Denmark, France, Japan, Sweden, Switzerland, West Germany, the United Kingdom and the United States. Of these, the cable noted, the

> French del took very strong position in favour of Article 23 (reco.), but maintained that it would not be wise to argue the point in the public session. He insisted that an argument on the issue in this forum would polarise the meeting and make much more likely the eventual adoption of the code under Article 21 (reg.). A number of other delegations agreed with the French position ... USdels continued to consult with these delegations, particularly Canada and France, on the appropriate way to handle this issue. Both Canada and France are represented on the Executive Board, where the issue will be formally considered next January.

The corridor lobbying and the informal discussions were effective, particularly on the opinions of WHO staff. The cable noted that the legal status was 'discussed at length' with them, and that a conversation at the end of the consultation with Dr David Tejada, assistant director-general of WHO, 'indicated that our point was understood'. The cable also said:

> All WHO officials stressed their desire for consensus on the code, and they recognize that proceeding under Article 21 would make consensus very difficult. The Secretariat left no doubt that it is under pressure from both sides, but it wants to find a formula acceptable to US.

The consultation went well. But as one Third World delegate told Baer, 'It went too well. It was too easy'. The implication was that neither the August nor September consultations were seen by the industry as important battlegrounds.

Baer explained that he thought the 'intense hard-core battle' would be fought at the WHO executive board in January 1981 and at the WHA in May 1981. 'One of the major dimensions of that battle will be the legal status of the code, and some of its key provisions.'

The analysis was right. The timing was not. And it was the failure to gauge accurately the extent of the backroom negotiations and trade-offs that were under way that left IBFAN in no real position to influence successfully either the content of the final draft or the code or its status.

Throughout 1980, IBFAN groups had been working most closely with those members of the WHO staff who were the strongest advocates of the code. It was a natural liaison, but in some ways unfortunate. For while they were offering clear advice on what would be best in an ideal situation, they were not always the people who made the political decisions within WHO—decisions which were based not so much on questions of infant health, but on the healthy survival of WHO.

With the United States paying for 25 per cent of WHO's regular budget and another dozen industrialized countries boosting that figure to 70 per cent, WHO had to listen very carefully to what the industrialized countries were saying. As the American cable pointed out: 'members of WHO staff went to considerable effort to seek out USdels for further discussion and clarification of US positions.'

The industry was also well versed in exploiting this type of economic dependence. Although spoken by a pharmaceutical company executive, the advice of Robert Dee, chairman of Smith Kline, was not far removed from the views of the baby-food industry.

> My second recommendation is that we put intense pressure on the UN—and on its health agencies—to give realistic instead of unrealistic advice to developing countries. For this year (1981) and next, 70% of WHO's budget will be paid by 13 industrialized countries—13 out of 156 WHO member countries. Certainly this entitles the industrialized world to stand up to WHO. We must have the will to do so.[29]

WHO and UNICEF began the last stage of the code development process immediately after the September consultation. The first version of the fourth draft of the code was ready within three weeks of that consultation. The United States had noted in its cable that it expected the fourth draft to be stronger than the third, and it was. However, the fourth draft itself went through four revisions! In the space of a week in late November, three revised drafts were produced.

An internal War on Want memo notes that each version became 'more unclear on the essential elements of the code, and an increasing amount of industry wording, and industrialized country wording found its way into the code.' None the less, it was considered by NGOs a vast improvement on earlier texts. IBFAN decided to do a line-by-line commentary on the fourth draft as a public document that could be circulated to members of WHO's executive board, in the belief that there would be opportunity for further changes.[30] That belief was wrong. On 10 December 1980, WHO sent out to governments the document that some five months later would be adopted as the International Code of Marketing of Breast-milk Substitutes. There were no further changes made to the text.

Not only was the content agreed, but by the end of November the deal was secured that finalized the legal status of the code. In a meeting with IBFAN representatives, Dr Tejada explained that, mainly as a result of its contacts with the United States' government, WHO was going to press for a recommendation. He said that the Scandinavians had somewhat reluctantly agreed with this line, provided that the code got a unanimous vote of acceptance in the WHA; was accompanied by a strong resolution; and the content was not watered down—that the present draft was to be considered a minimum. Tejada said that he was relying on the United States' government to ensure the other major industrialized countries gave their support for this proposal.[31]

Tejada also suggested that the NGOs should not mention this shift publicly, but rather should continue to press and argue for a regulation as that would be a way of countering industry pressure for a weakening of the code. He explained that the industry had heard of WHO's new preference for a recommendation a few days earlier, and had already launched an attack on the content of the code.

Both options—the regulation or the recommendation—would still be presented to the executive board, and also, Tejada thought, to the WHA, and it would be there that the final decision would be taken by the member states. He said there was always a possibility that if the deal with the Unites States fell through, the Scandinavians and some Third World countries would go back to pushing for a regulation.

With IBFAN barely a year old, and without enough experience in the intricacies of international lobbying and negotiation, Tejada's statements went basically unchallenged. It is interesting to speculate for a moment as to what would have happened if IBFAN had, at the end of 1980, denounced the compromise deal that had been arranged, exposed more fully the compromises in the code itself, and disassociated itself from the process.

It was a possibility that had been considered, although not too seriously. For small organizations that depended on mass participation and people's action, the high-level, detailed, almost esoteric, manoeuvrings of the code-drafting process were virtually impossible to communicate in an easily understandable manner. Discussions of whether to put an article controlling information before an article banning advertising in the code were not likely to capture people's interest.

During the year IBFAN had grown stronger—become more of a reality. It produced several reports documenting industry violations of the October 1979 meeting's recommendations.[32] It had gained the respect of several governments for its ability to argue clearly and concisely on this issue. It had found an ally in the international press who saw IBFAN as the best source of information on, and the logical counterbalance to, the industry. The focus on the code and the consultations and lobbying activities helped to strengthen the cohesiveness of the network. But it did so at the cost of local organizing and building the base support for some parts of the campaign. The old question of controlling the campaign was re-emerging. The code process had its own momentum, its own discipline, and, to a large extent, was beyond the control of the NGOs. They were a major part of the process, however, and, once involved, they had to follow the process through.

Although 1980 had some considerable successes for the NGOs, in retrospect it was not their year. The industry had not really won either, although on balance, despite their protests to the contrary, they had achieved a remarkable compromise. In Nestlé's words:

> . . . the final version of the code recognized that the industry did indeed have a role to play in Third World infant feeding . . . At last, industry had the endorsement it needed to justify the continued marketing of infant formula in developing countries![33]

However, 1981 was to be a different year.

Notes

1. *Business International*, 'Executive viewpoint', interview with Arthur Furer, president and chief executive officer of Nestlé, 1 June 1979.
2. K. Speck, 'Nestlé director: "We do not feel restricted"', *Tages Anzeiger*, 19 October 1979.
3. Ibid.
4. US Congress, *Marketing and Promotion of Infant Formula in Developing Countries: Hearings Before the Sub-committee on International Economic Policy and Trade of the Committee on Foreign Affairs of the House of Representatives*, Washington, DC, US Government Printing Office, 1980, p. 29.
5. Ibid., p. 74.
6. Ibid., p. 155.
7. Ibid., pp. 31–2.
8. Ibid., p. 96.
9. Abbott/Ross *et al.*, *World Health Organization Development of an International Code for the Marketing of Infant Formulas: Background and Perspective*, Columbus and New York, March 1981, p. 9.
10. R. Young, 'Food firms upset by UN stand on baby feeding', *The Times*, 6 March 1980.
11. ICCR, confidential notes from a conversation with an unnamed expert who attended the consultation, February 1980.
12. War on Want, notes on NGO consultation on the WHO/UNICEF code, March 1980.
13. WHO, internal briefing paper summarizing government responses to first draft of code for the marketing of breast-milk substitutes, Geneva, May 1980.
14. *INFACT Update*, 'Formula industry attempts to subvert WHO Code process', April 1980, p. 1.
15. Letter to Dr Halfdan Mahler from Dr John H. Bryant, Deputy Assistant Secretary for International Health, US Department of Health, Education and Welfare, 29 March 1980, pp. 1–2.

16. WHO, *Provisional Summary Record of the Seventh Meeting of Committee A, 33rd World Health Assembly* (Doc. A33/A/SR/7), Geneva, 16 May 1980, p. 2.
17. WHO, *Provisional Summary Record of the Eighth Meeting of Committee A, 33rd World Health Assembly* (Doc. A33/A/SR/8), Geneva, 17 May 1980, p. 5.
18. WHO, Doc. A33/A/SR/7, op. cit., p. 4.
19. The countries were Algeria, Burma, France, Ghana, India, Nigeria, Norway, Peru, the Soviet Union and the United States.
20. IBFAN, *A Capsule Summary of Events at the 33rd WHA* (mimeo), Geneva, May 1980.
21. A. Chetley, 'Pressure increases on infant food industry', *Outlook*, War on Want, No. 7, June/July 1980, p. 9. See also 'Bottling up the formula firms', *Development Forum*, June 1980.
22. Ibid., p. 10.
23. WHO, *Basic Documents*, Geneva, 32nd edn., 1982. The WHO Constitution, Articles 19 and 20 deal with conventions, 21 and 22 with regulations and 23 with recommendations, pp. 7–8. (It is worth noting that, since its founding in 1946, WHO has not attempted to introduce conventions, and has only introduced *two* regulations: one dealing with international scientific nomenclature for pharmaceutical and biological substances; and the other dealing with international immunization requirements, neither of which are controversial. The bulk of its policy on matters of health has been put forward as recommendations contained in resolutions. Thus governments which disagree with such recommendations can simply ignore them.)
24. Abbott/Ross *et al.*, op. cit., pp. 11, 14.
25. INFACT, *Notes on the Consultation of Third Draft of WHO/UNICEF Code of Marketing* (mimeo), Minneapolis, September 1980, p. 1. (One of the points that the industry stressed with WHO and UNICEF was that the consultation should be considered a 'private' meeting, i.e. that statements made by them in the meeting should not be publicized outside the meeting. This was mentioned at the beginning of the consultation, and repeated during the consultation, when the industry, looking towards the NGOs present, complained to the chair that it suspected that a journalist was present. The chair dutifully asked that if there were any journalists present they should leave. When no one left, he repeated the fact that it was a private meeting and discussion continued. The September issue of *IBFAN News* contained a front-page story, based largely upon comments made by the industry during the consultation. Within days of its publication, an irate Steven Bauer, vice-president of Wyeth, stormed into UNICEF's office in New York and gave vent to a long protest over the irresponsibility of NGOs. IBFAN subsequently apologized to WHO and UNICEF for any inconvenience the incident may have caused the two agencies, but felt that the action was justified in helping to demonstrate that the code development process was an open process, one that involved people, and that people had a right to know the real views of the industry.)
26. M. McComas, *Europe's Consumer Movement: Key Issues and Corporate Responses*, Geneva, Business International, December 1980, pp. I–93.
27. Taped interview with Edward Baer, 27 September 1980.
28. Ibid.
29. M. Muller, *The Health of Nations*, London, Faber & Faber, 1982, p. 179.
30. IBFAN, *Line by Line Commentary on the WHO/UNICEF International Code of Marketing of Breast-milk Substitutes*, Geneva, December 1980.
31. IBFAN, internal memo, 25 November 1980.
32. See, for example, *Infant Formula Promotion 1980*, Geneva, London and Minneapolis, IBFAN, May 1980 (this documents 331 violations of the October meeting recommendations which occurred between January and May 1980, involving twenty-one companies, in thirty-five countries); *Infant Formula Promotion: 100 New Violations*,

Geneva, IBFAN, September 1980 (this covers primarily violations which occurred between May and September 1980); *Baby Food Facts Foil Industry Attacks*, Geneva and Minneapolis, IBFAN, September 1980, (this counters a claim by ICIFI that all but fifteen of some 117 previously reported violations were invalid); *Infant Formula Promotion 1980*, Geneva, London and Minneapolis, IBFAN, January 1981 (this is a year-end compilation, documenting a total of 682 violations, involving thirty-four companies in fifty different countries).

33. M. McComas, G. Fookes and G. Taucher, *The Dilemma of Third World Nutrition: Nestlé and the Role of Infant Formula*, Vevey, Nestlé, 1983, p. 15.

7. Yes . . . and no

> The question which we should really ask ourselves is why do we want to adopt this code? And the response is unanimous: we want to adopt it in order to protect infants and young children. Now, let us be logically consistent and support this text unanimously.
>
> Dr L. Adende Manest, representative from Gabon,
> at WHO's executive board, January 1981

On 4 January 1981, the *Washington Post* ran a major article which implicated Ernest Lefever, the head of the conservative Ethics and Public Policy Center (EPPC), in an attempt by Nestlé to defeat the boycott and smear the groups involved.[1] Lefever had powerful friends in the new Reagan administration. He had served as a transition adviser to the president and apparently helped brief him on military and foreign policy in relation to Third World countries in order to prepare Reagan for his new place on the world stage. In return, later in the year, Reagan nominated Lefever as Assistant Secretary of State for Human Rights and Humanitarian Affairs.

Lefever, later in January, wrote an editorial piece for the *Wall Street Journal* which strongly criticized the proposed baby foods code as 'high restrictive'. The article showed little respect for the process by which the code had been developed. Lefever claimed that at the upcoming WHA,

> Third World delegates and their Soviet bloc friends may well adopt the code with little critical examination. Western governments will be under pressure to go along to keep the faltering North–South dialogue alive. This will be an unprecedented attempt at international legislation by ideological intimidation.

He concluded by suggesting that a 'new and more realistic administration' in the United States might well 'reject the compassion-coated revolutionary rhetoric and vote against the code'.[2]

The infant-food industry was quick to capitalize on Lefever's article. It and similar articles were circulated by ICIFI to members of WHO's executive board as they began their discussions on the code during January. The strategy backfired. Dr K. Ridings of Western Samoa singled out the 'noxious submission that comes from the Ethics and Public Policy Center in Washington', said that he resented the implications contained in it, and added that he thought that Dr Mahler and some of the WHO staff 'probably resent some of the scurrilous comments made about the Organization'.[3]

Mahler certainly did, and expressed some disappointment that the members of the executive board had not given the WHO staff more support for their efforts in drawing up the code in the face of such attacks. He deplored the characterization by Lefever and other right-wing commentators that WHO staff were secretive UN bureaucrats and said he felt that the kind of 'freedom of expression' these commentators had exercised allowed them 'the right to be as far from truth as one can be, including doing harm to children and the World Health Organization'.[4] Several delegations then spoke up in strong support of WHO's efforts.

ICIFI prefaced its material to the executive board with a letter from Ernest Saunders, the new president of ICIFI and a vice-president of Nestlé. In the letter, Saunders put the position of the industry very clearly: 'The World Industry has found this present draft code unacceptable . . . highly restrictive . . . irrelevant and unworkable . . . The various provisions, if applied, could have a negative effect on child health'.[5]

ICIFI held off circulating the letter until the executive board had decided upon another matter on the agenda—an application by ICIFI to enter into official NGO relations with WHO. It had been Dr Flache's first task as ICIFI secretary general to put in motion the process, and ICIFI had been confident that Flache's contacts and influence would secure the necessary support. Aside from the ability to address the various meetings of WHO, to circulate papers officially, and improved access to WHO documents, NGO status for ICIFI would have represented a WHO stamp of approval. It would have increased ICIFI's legitimacy enormously.

The idea that an industry lobby group could be classified as an NGO was not a new one. WHO already had accepted the International Federation of Pharmaceutical Manufacturers' Associations (IFPMA) and the World Federation of Proprietary Medicine Manufacturers as NGOs. With ICIFI, it was decided that the item should be deferred for a year to allow more time for working relations between the two organizations to be strengthened.

When the discussion on the code got under way, an interesting argument began to emerge. Several delegations noted that there were weaknesses in the code and suggested broadly that it might be an idea to amend some of the passages, although no specific proposals came forward. Dr A. Patterson of Jamaica noted that some of the definitions were 'very wide and anything can go through and some of them are so narrow that you can walk around all the edges'. Dr Ridings of Samoa said, 'the code is far too loose from the point of view of a developing country. In fact, it's got so many holes in it that an unscrupulous manufacturer could drive a herd of milk cows through it'.[6] Most delegations, however, were nervous about opening up the content of the code for detailed discussion. Certainly, a few loopholes might be closed, but in the haste necessitated by the lack of time at the executive board, new ones might be opened.

Dr Maureen Law of Canada picked up on these concerns and used them as a solid reason for introducing the code as a recommendation. She argued that a recommendation would be much easier to amend and modify in the light of experience.

> The step of adopting an international health regulation is certainly a very serious one and I think it is one that we should not undertake at this point in time unless we're very certain about the content ... Another alternative, I suppose, would be to postpone the consideration until the regulation has been made so perfect that everyone can accept it, but I think that no one wants to wait for action on this.[7]

Yet, despite all the possible problems, only a minority of countries represented on the thirty-member executive board actually suggested a recommendation as their first preference. In addition to Canada, only France, Switzerland, Jamaica and the Comoros indicated this. Nine countries—The Netherlands, Iran, Cape Verde, Congo, China, Gabon, Kuwait, Oman and the Soviet Union—were categorically in favour of a regulation. A further six—Mexico, Norway, Brazil, the United Kingdom, Turkey and Gambia—preferred a regulation but accepted that a unanimous recommendation would be better than a majority in favour of a regulation. Two countries—New Zealand and Burundi—expressed the hope that a recommendation would be tried for two or three years and then strengthened to a regulation, if it was seen to be necessary. The other eight countries on the board did not voice an opinion.

Dr J. Reid of the United Kingdom suggested that the 'very substantial moral power' which WHO possessed should not be underestimated. He said he felt that a code adopted as a unanimous recommendation would be 'a formidable instrument and one which would be more powerful than a less widely accepted regulation'. Dr T. Mork of Norway was less sure. He noted that since the first recommendation for action on this issue was passed by the WHA in 1974, few governments had done anything. At the same time, he noted that the companies had been aware since 1974 of the concern of the WHA, 'but they continued their regressive and often unethical sales promotion'. He suggested that the decision over the legal status depended on the answer to two questions:

> First, do we trust that the infant food industry, all governments, all health workers will pursue with all means possible the implementation of all provisions of the code? Second, does an adoption by consensus in the WHA of a draft code in the sense of a recommendation have the moral weight as the collective will of the member states to get the code implemented universally?

In answer to the first question, with regard to the industry, he said he had 'serious doubts based on statements saying they find the present draft unacceptable'. On the second question, he reserved his judgement, but it was fairly obvious that here, too, he had serious doubts.[8]

The United States, although not a member of that year's executive board, was present to observe the proceedings (and to lobby in the corridors). Dr Bryant was given permission to address the board, and gave what, in the light of later events at the WHA in May, appeared to be an unrealistic indication that the United States would in the end support a code. He noted the American reservations on a regulation, and said:

> We think it would be unfortunate if the WHO lost the ability to maintain the consensus on the vital issue of infant nutrition . . . In the light of all the agreement that exists within WHO and its membership on infant nutrition questions, it would be very unfortunate if the Assembly came to a conclusion on the matter through divisive action.[9]

It was as direct a statement as the United States was prepared to make about its willingness to support a recommendation in order to preserve unanimity. It was very clear that the United States would not support a regulation.

The executive board approved the code, in its entirety, with no changes, and forwarded it to the WHA with a resolution calling for its adoption as a recommendation. There was, however, a tacit understanding which emerged throughout the discussion and which was implied in the resolution, that strong efforts would be made to monitor the effectiveness of the code, report back in 1983 and, if necessary, change the code to a regulation.

It seemed to be the most reasonable approach. It would serve warning on the industry that it had better behave; it would give governments a chance to try the code in practice and come back with ideas for amendments; and if it was seen to be ineffective in controlling marketing abuses, there would be no excuses or arguments left that would prevent a regulation. It was a very flawed analysis, as subsequent events proved.

Battling for position

After the executive board meeting, efforts were made at international and national levels by the industry, the United States' government and the NGOs to try to get a sense of how countries planned to vote in May. By March, the scene was set for the coming assembly.

The first major event concerned a Swedish paediatrician, Dr Stig Sjolin. After twenty-three years as a consultant to Nestlé's subsidiary company, Findus, in Sweden, Dr Sjolin resigned in protest over Nestlé's baby-food marketing practices. In a letter to *The Lancet*, Dr Sjolin said,

> I once believed that I would be able to influence Nestlé's marketing practices . . . this hope has now evaporated . . . Nestlé, in its obstinancy, has even joined with other large manufacturers of breast milk substitutes—not, as is now clear, to establish rules of conduct that would protect child health, but to defend the manufacturers' activities and to protect their own economic interests.[10]

Sjolin urged his fellow paediatricians to take a firm stand on this issue and to 'mount a concerted campaign to convince the powerful infant food industries' to accept the WHO code.

The infant-food companies had other ideas. Although there was 'consensus' amongst ICIFI companies that they should accept the code and work for a weak resolution implementing it, there was not unanimity, according to Geoffrey Fookes of Nestlé.[11] In discussing the situation with a representative from the American mission in Geneva, Fookes said 'there were the "hawks" in the international group who truly desired to see a confrontation at the upcoming WHA'.

What the 'hawks' hoped was that if the United States spoke out against the code, the assembly would react by adopting it as a regulation. At that point, other industrialized countries would also speak out against the code. Fookes added that the industry had 'indications that either Canada or Australia or the Federal Republic of Germany would then take the WHO to the International Court of Justice on procedural irregularities in the development of a regulation'. The particular attack would be over the fact that the code had not been developed through intergovernmental negotiations. In this way, the industry 'hawks' believed that both 'WHO and the regulatory code would be thoroughly discredited'.[12]

Fookes also told the American representative that the industry was 'not really worried that any countries will in fact try to implement the code "in its entirety" or "as a minimum requirement"'. It was an indication of the industry's confidence that in bilateral negotiations between the industry and individual countries, the provisions of the code that would most restrict marketing practices could be satisfactorily weakened. It was also a confirmation of the concerns expressed by Dr Mork of Norway at the executive board. Fookes said the industry's main concern about the code related to its relations with pressure groups critical of the industry. He saw that unless it was 'crystal clear' that there was a need for flexibility in implementing the code, the 'industry would be held to the highest apparent standards' even though some countries might choose less strict criteria.

One of ICIFI's 'hawks' was American-based Wyeth. Together with the other two major American companies—Abbott/Ross and Bristol-Myers—they launched an all-out assault on the United States' government during March. They issued several thick position papers on the code and began a two-week series of visits to officials of all affected federal agencies and key legislators to urge the United States to vote against adoption of the code.[13]

The industry lobby prompted a flurry of activity in both the United States' State Department and the Department of Health and Human Services. A cable went out under the signature of Secretary of State Alexander Haig to American embassies in all industrialized countries requesting them to 'consult all relevant ministries in host government—foreign affairs, foreign assistance, trade, economic affairs, justice, health, etc.—to advise on positions governments plan to take', and also to report back on how other governments would react to four possible positions the United States might take at the WHA:

(A) If US voted no, would others vote no, or would others abstain?

(B) If US abstained, would others abstain?

(C) If US sought to create a drafting group and change the text of the code, would others support such a move? What kinds of changes would they want to introduce themselves?
(D) If US sought to create a drafting group or otherwise to change the text of the implementing resolution, would others support such a move?[14]

According to the *Washington Post*, there was a clear indication back to the State Department that other Western nations would 'refuse to go along with any US opposition to the code'.[15] This was confirmed in a memo to the Secretary for Health and Human Services from Dr John Bryant, Deputy Assistant Secretary for International Health, which said there were 'particularly strong feelings among European and other developing countries that any weakening of the code is unacceptable'.[16]

Bryant prepared a second memo which examined some of the questions that had been posed by the industry.[17] He found that a Canadian study on the impact of providing infant formula samples to mothers appeared 'to contradict industry's position that there are no scientific studies that show a cause and effect relationship between marketing practices and declines in breast feeding'. A study from Papua New Guinea showed that the introduction of strong legislative controls over marketing had a positive result in increasing breast feeding and in improving infant health. A study from Brazil revealed that there was a clear association between advertising and promotion of infant-feeding products via health workers and high rates of bottle feeding and infant malnutrition.

A tense assembly

By the time the WHA opened on 4 May 1981, the American delegation still had no firm instructions on the position it should take on the code. The industry had no such problem. The *International Herald Tribune* quoted Stanislas Flache of ICIFI as saying 'We oppose the universal code and some believe it is a sign that the UN system is moving to control multinationals'.[18]

ICIFI followed this up by opening a 'Press and Information Centre' in a suite at the expensive Inter-Continental Hotel, only a few minutes away from the Palais des Nations in Geneva where the WHA was being held. The 'information centre' was designed to provide journalists and delegates with materials about the code, about ICIFI, and about the effectiveness of voluntary industry codes already in existence, and, therefore, the lack of necessity of an international code.

One of the documents was a 'Charter for the provision of safe and adequate nutrition for infants'. It contained, on the front cover, a quotation from Dr H. Yellowlees, the Chief Medical Officer of the British Department of Health and Social Security. The implication of the use of the quote was that the United Kingdom was supportive of ICIFI's position. NGOs drew this to the attention of the British delegation at the WHA. The delegation was embarrassed and somewhat annoyed, and the following day circulated a letter to all delegations at the WHA noting:

> The reproduction of this quotation may create the impression that the UK Government or one of its officials has some connection with the International Council of Infant Food Industries. I wish to make it clear that this reproduction has been done without any authorisation or agreement by or on behalf of the original signatory or of the UK Government.[19]

The UK also noted that it was firmly in support of the code and the proposed resolution to adopt it. The timing could not have been better. The indecision of the United States had begun to affect other delegations at the WHA. Although several governments made strong declarations of support in general plenary sessions, many more countries were beginning to think that they should abstain when the vote came up on the code.[20] The British statement was seen as representative of the positions of other Western European countries and helped ease some worries other countries had about whether the code would be approved.

By the end of the second week of the assembly, the air had cleared completely. News leaked out of Washington that the United States would definitely vote against the code, although the delegation in Geneva maintained the final position was still to come. As the *Financial Times* reported:

> . . . the decision to vote against the World Health Organisation's Code was taken by the White House, apparently by Mr Edwin Meese, the President's right hand, after direct representation by industry and over the more cautious advice of lower level bureaucrats in both the State and Health and Human Services Departments that the US simply abstain.[21]

Protest in the United States

The United States' government stand and the two votes on 20 and 21 May triggered off an avalanche of editorials expressing dismay in newspapers around the United States, from the *New York Times* and the *Washington Post* to the *Buck County Courier Times*. Within a three-week period there were so many editorials written that INFACT was able to compile a bound book of them one and a half inches thick.[22]

The protest did not stop there. Two officials from USAID, Dr Steven Joseph, a paediatrician and the chief health professional at AID, and Eugene Babb, deputy assistant administrator for food and nutrition, called a press conference threatening to resign if the United States voted against the code. After the vote, explaining his resignation, Joseph said he found the American position,

> unconscionable, inimical to the responsibilities of health professionals, contrary to the best interests of my country, inexplicable to my professional colleagues from developing and developed countries alike and, most important, damaging to the health and growth of the world's children.[23]

Babb called it a 'wrong decision from every conceivable perspective'. He said voting 'no' on the code would 'signal the world that the US does not care if our firms behave responsibly overseas'.[24]

Babb and Joseph were then supported by the calling of four different Congressional hearings to examine the American vote. Public demonstrations were held in five major American cities. More than 10,000 letters and telegrams poured into the White House and the State Department. The United States' House of Representatives on 16 June, by 301 votes to 100, approved a resolution expressing its dismay at the vote, while the Republican-controlled Senate on 18 June voted 89–2 in favour of a similar resolution.[25]

By an ironic coincidence, on the same day that Joseph and Babb announced their resignations, Congressional hearings opened to decide the nomination of Ernest Lefever as Assistant Secretary of State for Human Rights and Humanitarian Affairs. Lefever probably would have had a difficult time at the hearings, even without his involvement in the baby-foods issue. His views on human rights were founded on a strong anti-communist bias. His view was quite clear that 'the Soviet Union is not only the greatest violater of human rights but the greatest disturber of peace in the world'. At the same time he suggested that the United States should sell military hardware to South Africa. His views on Africa in general were far from diplomatic. 'If it took the barbarian tribes of Europe more than 1,000 years to emerge from [*sic*] nationhood, how long will it take these equally primitive tribes?' He felt that the United States had 'no right and no responsibility to express its concern, either publicly or privately, for the human rights of people in other countries'; that efforts by the United States' government to influence the behaviour of other governments towards their own citizens were 'violations of national soveriegnty', and that the United States should not impose any human rights conditions on its aid, trade or other foreign relations.[26]

His position over the baby-milk code did not help him. Lefever said 'I support the President of the United States in this vote, in the name of humanity, human rights and babies in the Third World'.[27] His whole handling of the questioning about his contacts with Nestlé led Senator Alan Cranston to comment that Lefever's conduct was

> highly questionable. Not only did one aspect of this situation violate the [Ethics and Public Policy] Center's own ethical standard to refuse any contribution from a donor for a project in which the donor had a direct interest, but Mr Lefever's entire handling of this study and the three contributions in question is at the very least an apparent conflict of interest to which Mr Lefever is demonstrably insensitive.[28]

The Senate committee voted 13–4 against Lefever's nomination on 5 June. Before the decision went to the whole Senate for a final vote, Lefever withdrew his name as a candidate for the position.

The lonely vote

With the situation in Washington heating up, one Third World delegate at the WHA described industry lobbying against the code as 'the only contamination in the otherwise clinical atmosphere of Geneva'.[29] The efforts undertaken by the industry were almost unbelievable. Three days before the code came up for discussion, a previously unheard-of-organization called Voices of the World's Children (VWC) circulated to the press and to delegates three documents condemning the code and claiming that the International Paediatrics Association (IPA) was also opposed to the code. Fortunately, two members of the IPA's executive were delegates to the WHA and publicly repudiated these claims. Enquiries by Geneva journalists as to who was behind the VWC found that it had just been formed, that no one knew who was funding it, and that the only source of information about it came from the ICIFI Press and Information Office. VWC vanished without trace after the code was approved.[30]

Another unknown organization, the All India Medical Students' Association, which claimed to be affiliated to the International Federation of Medical Students' Associations, used a known pro-industry lobbyist to circulate a letter to delegates implying that Indira Gandhi did not support the code. This prompted the Indian delegation to set the record straight in a letter to WHA delegates. The Indian delegation also threw grave doubts on the validity of either the organization or its backers.[31]

Antonio Carrera spent much of the assembly sitting at the table reserved for the Guatemalan delegation. It was not until two IBFAN representatives accidentally overheard a conversation he was having, where he outlined a strategy to confuse debate on the code by introducing a long series of amendments, that some investigations began. A preliminary check with WHO officials did not turn up his name as an official delegate. Private enquiries were made by both WHO and IBFAN with the Guatemalan delegation, and shortly before WHO took official moves to have him ejected, Carrera left the assembly. He was a Guatemalan lawyer who, from time to time, was reported as having done work for Nestlé in Guatemala.[32]

ICIFI also used a series of 'carrots and sticks' with delegates in an attempt to influence their votes. Several delegations reported offers of lavish meals, free drinks and sightseeing trips around Switzerland, while any delegate who was critical of the industry quickly received a letter from ICIFI suggesting a little talk. Thus, when Dr M. Karpeh, the deputy Minister of Health of Liberia, said during the WHA,

In recent times we have been observing a marked increase in the number of cases of diarrhoea among infants due to increased artificial feeding of babies as a result of brain-washing the population through constant advertisement by the producers. It is a known fact that in most developing countries, besides the adverse effect of artificial feeding, the average family cannot afford the high cost. This is just as detrimental as providing dangerous drugs to the populace . . .[33]

he soon received a letter from ICIFI secretary general Flache asking him to provide details of any promotional activities and copies of any government studies that showed links between increased diarrhoea and the unnecessary use of infant formula. Flache also suggested a meeting to discuss the situation further.[34]

When the code finally came up for discussion at the WHA, it was almost anti-climatic. Following the usual procedure, one of the two major committees at the assembly had the issue on its agenda. The committee would make a recommendation which would then go before the full plenary session for final adoption. On 20 May, more than fifty countries signified that they wished to speak on the code in the committee. With an eye on the clock, the WHO secretariat hoped to finish debate on the code within a single afternoon. It was arranged that the Belgian delegate would move closure of debate shortly before the end of the day's business. By that time, most delegations should have had a chance to speak. However, signals became crossed, and after only nine countries had spoken, the motion for closure came. Confusion and some anger erupted. Some delgates were unsure whether they were voting for closure or for the code. As a result both Chad and Bangladesh mistakenly voted against the code. There were no mistakes the following morning in the plenary session: 118 countries said 'yes' to the code; three countries—Argentina, Japan and South Korea—abstained; and only one country, the United States, said 'no'.

It was left to Gerald Helman, the American ambassador to the UN in Geneva, to try to explain the vote. He said the code contained

provisions that will cause serious legal and constitutional problems for the US itself . . . We also have serious concerns about this Organization's involvement in commercial codes, and this is a central basis for our inability to support this code.[35]

Despite this explanation, many delegates at the WHA found the American position unpalatable. Dr A. Al-Saif, the deputy head of the Division of Preventive Medicine at the Kuwait Ministry of Health, summed up the views of many others when he said in the plenary session,

This code is a landmark in the history of health, in particular the health of children, particularly for the developing countries. Today, we believe that this is our minimum duty in adopting this code, in view of the needs of protecting health for children. We therefore wonder why certain countries, particularly in the case of the major powers, are adopting a position against this international code, in order to hamper our work. We think this is a deplorable and regrettable situation which will be recorded in the annals of history.[36]

Notes

1. M. Mintz, 'Infant formula maker battles boycotters by painting them Red', *Washington Post*, 4 January 1981.
2. E. W. Lefever, 'Politics and baby formula in the Third World', *Wall Street Journal*, 14 January 1981. (Lefever had long been critical of the World Council of Churches (WCC) and its various national affiliates for their attitudes towards development and their support of policies designed to assist the poor and powerless overcome the economic, social and political forces which maintained this situation. He saw the infant-formula controversy as being masterminded by the WCC, the National Council of Churches (NCC) in the United States and the British Council of Churches (BCC) in the United Kingdom. He claimed in the article that these church councils were using the issue as a way of attacking multinationals, and in the process they failed to display respect for 'medical evidence, reasoned argument, fair play or simple honesty'.

   The same charges could more accurately be levelled at Lefever over the content of this article. First of all, the WCC has, at best, been marginally involved in the campaign only through its Christian Medical Commission. Most of its involvement has come through efforts in the field—advising health workers to avoid the use of feeding bottles—rather than in public campaigning. The NCC in the United States has had a somewhat more visible profile through its endorsement of the Nestlé boycott and its support for ICCR, but it could not be considered as one of the policy-making bodies involved in the planning of the international campaign. The BCC has said and done virtually nothing on this campaign. Yet Lefever, whose Ethics and Public Policy Center was supposed to be 'studying' the controversy, claimed that *The Baby Killer*, War on Want's report, was issued by 'a periodical affiliated with the British Council of Churches'! War on Want is not and never has been affiliated to the BCC.)
3. Unofficial transcript of tapes of the executive board meeting, 28 January 1981.
4. Ibid.
5. E. W. Saunders, letter from ICIFI to executive board members, January 1981.
6. Unofficial transcript, op. cit.
7. Ibid.
8. Ibid.
9. Ibid.
10. S. Sjolin, 'Should paediatricians support infants and/or the baby food industry?', *Lancet*, 14 March 1981.
11. Cable from the US mission in Geneva to the US State Department, Washington, DC, March 1981.
12. Ibid. (When the contents of the cable were made public by NGOs during the May 1981 WHA, the head of the Canadian delegation stated that 'the Government of Canada is not contemplating such an action now or in the future'.)
13. M. Mintz, 'Baby formula producers challenge UN agencies', *Washington Post*, 18 March, 1981. (The industry materials were *The Weakness of the Scientific Premises Underlying the WHO Code for Marketing of Breastmilk Substitutes*, and *World Health Organization Development of an International Code for the Marketing of Infant Formulas: Background and Perspectives*.)

14. A. Haig, US State Department cable to all American embassies in OECD countries, March 1981.
15. M. Mintz, 'Administration urged to back strict code on breast-milk substitutes', *Washington Post*, 3 May 1981.
16. Ibid.
17. J. H. Bryant, 'Studies on relationships between use of and/or marketing of infant formula and breast feeding', memo to members of the (US) Interagency Task Force on Marketing of Breastmilk Substitutes, Washington, DC, Department of Health and Human Services, 9 April 1981.
18. A. Krause, 'World Health agency to weigh code on baby food', *International Herald Tribune*, 4 May 1981.
19. WHO, *Communication from the Delegation of the United Kingdom*, Doc. A34/INF.DOC./10, Geneva, 14 May 1981.
20. IBFAN, *Quotations by National Governments on WHO/UNICEF Code of Marketing and WHO Executive Board Resolution*, Geneva, May 1981.
21. J. Martin, 'Passionate politics in a milk-and-water affair', *Financial Times*, 22 May 1981.
22. INFACT, *Editorials and Opinions*, Minneapolis, 1981.
23. Statement of Dr Stephen C. Joseph in opposition to the US government position on the WHO/UNICEF International Code for the Marketing of Breast-milk Substitutes (mimeo), issued to the press, 18 May 1981.
24. N. Ram, 'Code on infant formulas: US decision revolting', the *Hindu*, Madras, India, 21 May 1981.
25. INFACT, 'Irked citizens say "No" to US vote against WHO baby code', and 'Congress condemns US negative vote on WHO Code', *INFACT Update*, Spring 1981, pp. 1, 6 (See also *US Congressional Record*, No. 91, 16 June 1981, pp. H2944–H2963; No. 92, 17 June 1981, pp. S6380–S6383, S6389–S6390; No. 93, 18 June 1981, pp. S6484–S6495, for text of Congressional resolutions, comments by Senators and Representatives, and additional material, including extracts from editorials submitted for the record.)
26. US Senate, *Hearings before the Committee on Foreign Relations of the United States Senate on the Nomination of Ernest W. Lefever to be Assistant Secretary of State for Human Rights and Humanitarian Affairs*, 18, 19 May, 4, 5 June 1981, Washington, DC, US Government Printing Office, pp. 97, 112, 504, 506.
27. Ibid., p. 362.
28. Ibid., p. 517.
29. Oxfam, internal memo, 10 June 1981, p. 4; and WHO, Doc. A34/INF.Doc./12, May 1981.
30. A. Chetley, 'Breast is best: the World Health Organization's International Code of Marketing of Breast-milk Substitutes', *IFDA Dossier*, No. 25, September/October 1981, p. 45. See also A. McGregor, '157 nations to get baby food code', *Times*, 22 May 1981.
31. Oxfam, op. cit., p. 2.
32. Ibid., p. 3.
33. IBFAN, op. cit.
34. ICIFI, letter from S. Flache to Dr M. Karpeh, 19 May 1981.
35. D. Munyakho, 'War on breast milk substitutes far from over', *Sunday Nation* (Kenya), 24 May 1981; and 'Battle of the Bottle', *Asiaweek*, 19 June 1981.
36. A. Chetley, op. cit., p. 42.

8. Bringing the code home

> This should be a celebration of commencement—of where to go from here—as well as congratulations to those of you who have laid the groundwork that brings us to where we are today. We are here because of the solid and persistent work, at times in the face of criticism and harassment, that has been carried out over many years.
>
> Dr Stephen Joseph, at INFACT's 5th National Conference, 10 July 1981.

Jubilation and celebration were the feelings that swept through the NGO network on 21 May 1981. A major victory had been won. An IBFAN press statement put out after the code was adopted listed five elements of the victory that were historic firsts:

1. An international code to restrict multinational business has been adopted by the members nations of a UN agency. The purpose of the code is to protect health.
2. The First and Third Worlds were in near-unanimous agreement on the need to regulate multinational industry and on a specific mechanism to do so.
3. Non-governmental organizations and consumer groups played a leadership role in calling for the code, in participating in the development process, and assuring its adoption. Now NGOs are specifically called upon to help implement and monitor the code.
4. There is explicit international consensus that the activities of some multinational corporations are responsible for ill health, and without strict controls there can be no enduring solution to infant malnutrition and death.
5. . . . proposals for appropriate control of marketing activities will be applied in both developed and developing regions of the world. There will be no double standard on infant health.[1]

At the same time, there was clear recognition that much more needed to be done. Dr Raj Anand, speaking on behalf of IBFAN, said, 'To relax our vigilance now would be foolhardy. Our babies are our future and they deserve all the gentleness and protection we can provide'.[2]

In under two years since its founding, IBFAN had become a major force in the infant feeding issue. The *Guardian*'s Iain Guest said of IBFAN in early 1981:

> It runs on a shoestring budget ($46,000 this year), but it has concentrated the energies of various pressure groups (including Britain's War on Want) in a quite remarkable fashion. IBFAN spokesmen[/women] regularly meet counterparts from industry on its own ground, and leave them floundering.[3]

IBFAN's strength came from two sources: the clear, single focus at the international level of pressing for the adoption of as strong a code as possible; and the diversity of skills, experience, contacts and resources that the various affiliates of IBFAN could contribute to the campaign. The single focus meant that every group, whether in India or Ireland, had a common objective and a common strategy: generate popular support for the code and exert pressure on policy-makers to vote for the code. The tactics were different in different countries and regions, but the results were largely the same.

In Asia, for example, during April 1984 an international media tour to generate support for the code was organized. The International Organization of Consumers' Unions (IOCU) coordinated the arrangements and provided a staff person and some funding for the tour. Local consumer groups in Bangladesh, India, Malaysia, Hong Kong and the Philippines set up press conferences and tours of health-care facilities and the communities. Dr Raj Anand, a paediatrician from Bombay and medical adviser to the Consumer Guidance Society of India, volunteered his time and experience to take part in the tour. INFACT asked one of its supporters—actress Linda Kelsey, star of the popular TV series *Lou Grant*—if she would join the tour. War on Want provided background materials for the tour—a complete information pack for journalists. It was a prime example of international cooperation, and was a tremendous success, attracting headlines and radio and television interviews in all the countries, except the Philippines. There the coverage was not as good, as the local Nestlé subsidiary went to great lengths to point out to newspapers and other media how much the company spent in advertising with them.[4]

Upon her return to the United States, Linda Kelsey told American journalists,

> I'm very angry. I have witnessed a situation that is very grave and, from what I learned, completely avoidable . . . I was very moved by the sense of helplessness that an underfunded health worker feels when faced with the powerful interests of large commercial firms such as Nestlé. They expressed the need for the Code to help right the imbalance, and help them attain the political force needed to protect their children.[5]

In the United Kingdom, a different approach was used by IBFAN affiliates. Beginning in late 1980, War on Want and Oxfam, assisted by the Baby Milk Action Coalition (BMAC) and Third World First (3W1), started mobilizing popular and political support for the code. A briefing session was arranged at the House of Commons for Members of Parliament (MPs), and shortly afterwards, an Early Day Motion (EDM) was put forward calling on the government to vote in favour of the code.[6] Individuals around the country were asked to contact their MPs, encouraging them to sign the EDM. It received the support of over 200 MPs from all parties, and led to an adjournment debate on the issue, followed by a deputation from MPs and NGOs to the Parliamentary Under Secretary for Health, Sir George Young. Sir George agreed the code was important, that it should be supported, and that the British delegation to the WHA 'would be briefed accordingly'.[7]

Edward Baer's description of the way an IBFAN team handled lobbying at code consultations or at the WHA could equally apply to the activity of the network as a whole.

> One of the things that we see shaping up in these Geneva operations is that different people have different concerns about different aspects of the whole operation. One person is a master expert at documentation. Another makes us think about how we are going to consult with the national governments, how we are going to contact them, lobby them and get our documentation into their hands. It's an interesting division of labour that takes place on a very informal, if not unconscious, basis. There is a certain kind of magic and chemistry that takes place when people from all over the world fly in and we begin to work together.[8]

Thus different organizations in the network performed different functions that benefited the entire network. IOCU concentrated on developing contacts with consumer organizations and other groups, particularly in developing countries, to help expand the network. The Geneva Infant Feeding Association (GIFA), a support group formed in 1979, maintained contact with WHO and other Geneva-based international agencies. War on Want helped supply the expertise to produce publications and campaign materials. Breast-feeding promotion groups in Sweden, Norway, Kenya and Trinidad helped to generate good information to help mothers breast feed. ICCR maintained contact with UNICEF in New York and began to develop a catalogue of scientific evidence about the issue. INFACT maintained an international mailing list and produced *IBFAN News* on a regular basis. All these groups, plus dozens of others around the world, conducted regular monitoring investigations to provide up-to-date evidence of company activities.

However, with the passage of the code, the major international focus began to disappear. As Baer noted during a press conference at the May 1981 WHA,

> At this point in time, the WHA is the one single focal point for all the activity, interest and pressure. After the end of next week, everything is going to disperse and at 156 national government levels, there will be opportunity for a negotiating and bargaining process to take place. Unfortunately, the infant formula industry has the interest and the resources to work very hard at that national level to see that codes are adopted that are weaker than the WHO/UNICEF Code.[9]

IBFAN, too, recognized the need to stimulate more action at the national level. A first step in that direction was taken immediately after the 1981 WHA when IBFAN held its first congress, which involved eighty-eight participants from thirty-nine countries. The purpose of the congress was to facilitate a better exchange of information about the activities and experiences of different groups, to plan future strategies for joint action, including deciding what IBFAN could and should do, and to begin to establish a democratic decision-making process. The congress approved a basic commitment to strengthening national groups and established five aims and objectives:

1. To promote maternal and infant health by encouraging and protecting breast feeding and other appropriate infant feeding practices.
2. To work for the end of promotion of breast milk substitutes and other products which interfere with breast feeding. The WHO Code and its accompanying resolution provide the minimum framework for action at present.
3. To develop, expand and strengthen links among groups and individuals pursuing these objectives.
4. To analyse the problem of maternal and infant health within the broader context of political, economic and social structures.
5. To establish relationships with other international organizations and networks in pursuance of these objectives.[10]

Specific activities to implement these objectives included improved monitoring of company promotional practices, pressure for national legislation based on the code, the development and dissemination of pro-breast-feeding information to counter the misinformation put out by the industry, and mobilization of community organizations, especially women's groups, to focus on the necessary changes in hospital routines and practices that would meet women's needs and promote breast feeding.

Six regional representatives were elected, one each from Asia, Africa, Latin America, North America, Europe and the Caribbean, and together with representatives from five key organizations—ICCR, INFACT, IOCU, GIFA and War on Want—were to undertake the functional coordination of the network. Over the next three years, this structure was further refined. In February 1982, the IBFAN Coordinating Council (IBCoCo) was expanded with the subdivision of some of the regions. Three 'central offices' were established, in Geneva, Minneapolis and Penang. Each office became the primary link for four regions, providing back-up, technical support and assistance with fundraising. The Penang office worked with the South Pacific, Far East, South East Asia, and South Central Asia regions; the Geneva office with the Middle East, East and Southern Africa, West Africa and Europe; and the Minneapolis office with North America, the Caribbean, Central America and South America. In 1984, IBFAN hired its first general administrator, who is based in London, to provide further coordination and extra support on fundraising.

A UNICEF observer commented in 1982 that the 'growing pains' IBFAN was going through represented

> a fascinating experiment in inter-NGO cooperation. First, the network includes a great diversity of groups working at several different levels, linked primarily

> by their interest in one substantive area . . . Second, IBFAN demonstrates the problems, as well as the potential effectiveness, of stimulating local and community activities through international planning and coordination.[11]

With limited financial resources, no staff initially (GIFA, INFACT and IOCU 'loaned' staff to IBFAN to perform the central office functions), and a general public feeling that with the passage of the code, the problem was 'resolved', IBFAN faced a daunting task. This was so particularly in industrialized countries where the impact of inappropriate marketing was less obvious and maintaining support for the campaign was difficult. Yet it was from the industrialized countries that most of the financial resources for the campaign were available. In developing countries the financial constraints existed, but a far more fundamental problem was the lack of a strong tradition of NGO action and, in some countries, an outright official suspicion of their activities.

IBFAN also faced the structural problem that communication and information flows within the world system are biased in favour of the industrialized countries. Thus, it is easier to communicate between London and Tokyo than it is between Nigeria and Kenya. Similarly, information tends to move rapidly from developing countries to the industrialized countries, but much more slowly in the reverse direction, particularly if it is information that might affect economic matters such as trade.

IBFAN could not institute a major restructuring of global communications systems. But it could, and did, implement a small-scale alternative approach. The network had been founded as a result of the personal contact, trust and understanding of a few individuals from different organizations, and it was this foundation that was slowly being expanded. A series of training, skills development and information-sharing seminars was arranged in different countries around the world during 1982 and 1983. Each seminar brought together people from groups and organizations in the region, usually comprising fifteen to twenty people, plus three or four members of the IBCoCo from outside the region. Particular emphasis was placed on ensuring that South–South links were strengthened. Thus a seminar in the Philippines included representation from South America, while one in the Caribbean included representation from Africa.

The result of each of the seminars was the establishment of a stronger regional network of groups and the development of a plan of action for the region as a whole. The cross-fertilization amongst different regions led to better understanding of the problems faced and an exchange of ideas about how to resolve them. The regional plans were then fed back through the IBCoCo and similarities extracted as a guideline as to where international coordination could be useful in facilitating local activities. In the process the regional coordinators have slowly been asserting more of a leadership role within the network. It is anticipated that this will continue in the future as more functions are decentralized. Both through the seminars, national activities and improved media attention and the wide dissemination of an 'Action Pack' on the issue, the number of groups affiliated to IBFAN increased to over 100 in some sixty countries by the middle of 1983.[12] At the same time, the effectiveness of the campaign increased.

National action

In India, when the government set up a committee to draft a national code, the infant-food industry set to work to influence the process. Nestlé's subsidiary in India offered a member of the committee, who was president of the Indian Academy of Paediatrics (IAP), a 'donation' of approximately $7,000. Shortly after the offer was made, this person presented to the committee a draft code written by the industry. The Consumer Guidance Society of India uncovered the information about the offer of money, identified the source of the draft code and lodged a strong protest with both the IAP and the Indian government. The government threw out the industry code, took the WHO model and strengthened it, and in late 1983 published it in the official law gazette as a prelude to its coming into force as a new law. The IAP voted overwhelmingly to refuse the Nestlé offer of money.[13]

This incident had a long-term effect, for in 1983 the IAP decided to refuse all baby-food company offers of support for its annual conference. In previous years, the industry had been a major contributor. In 1980, for example, Nestlé had spent approximately $20,000 on a dinner for delegates and other guests at the annual conference.[14]

In Kenya, the Breastfeeding Information Group (BIG) has been working for many years, providing information, a mother-to-mother counselling service, talks in clinics and advice and internal training for health workers. It also established an international committee to maintain links with organizations and groups, including IBFAN, that operated outside Kenya.

As a profitable market area, Kenya was a major focus for the infant-food industry. In April 1981, the Kenyan government announced a voluntary code of ethics for the marketing of infant-formula products. The code had been written by the industry, who naturally greeted its adoption with satisfaction. Health workers and NGOs, including BIG, were not so pleased and started a campaign to have the code withdrawn. By 1982, the pressure was beginning to pay off. The Kenyan government began drafting a new code, based more closely on the WHO code. This time, the writing of the code was not left to the industry—women's organizations, health workers and NGOs like BIG were all involved. By April 1983, a draft was finalized which, although not quite up to the standard of the WHO code, was a vast improvement on the industry code.[15]

In industrialized countries, most NGOs at first became involved in the infant-feeding issue because of the harm that was occurring to infants in developing countries. They used the issue as an effective illustration of the way in which decisions made in industrialized countries—either by governments or companies—can affect the health of people in developing countries. Campaigns were organized to put pressure, usually on companies, to change practices.

The Nestlé boycott is the biggest example. However, smaller actions have also been effective. For several years, a coalition of groups in The Netherlands had been active at taking Dutch companies to task over their marketing practices in developing countries. One of the members of the coalition, the Workgroup on Baby Food and the Third World, decided in early 1983 to launch a three-month campaign to focus on one Dutch company—Cooperative Condensfabriek Friesland (CCF). CCF manufactured infant formula and evaporated and condensed milk as well as other milk drinks under the brand name 'Dutch Baby'. The label of most of these products showed a picture of a woman dressed in traditional Dutch costume holding a baby. The Workgroup called on CCF to change the brand name and remove the picture, claiming that it could mislead consumers into thinking all the milk products were suitable for infants. The three-month campaign achieved considerable public attention in The Netherlands and was due to culminate in a national day of action on 27 April 1983. A few days before the day of action, CCF gave in to the demands and announced that it would discontinue using the picture and the Dutch Baby brand name.[16]

In developing the CCF campaign, the Workgroup liaised closely with consumer groups in Malaysia, one of CCF's major markets. The Malaysian groups provided information and examples of labels, and also put pressure on the company's subsidiary. It was a good illustration of parts of the international network cooperating in an activity.

Another example of the way the international network operates effectively came from Nigeria. In May 1981, Oxfam received information from a contact that a product called Daily Baby Food Custard Powder was being advertised as suitable for infants. The advertisements read, 'for strong growth of your baby, insist on nothing but Daily Baby Food Custard Powder'. The custard powder was manufactured by a British company, Bestoval, which exported it to a Nigerian company, New Century Foods. Oxfam's first step was to write to both companies asking them to alter their marketing practices. This resulted in the British company threatening Oxfam with legal action, followed by, after an exchange of correspondence, a disclaimer: 'We only make the stuff, it's up to the Nigerian company how they sell it'. The Nigerian company never replied to Oxfam. Finally, Oxfam asked the Nigerian government to look into the matter. Within weeks, the Nigerian Ministry of Health ordered the removal from the market of this and another brand of custard powder using similar sales tactics. Food inspectors were instructed to seize existing stocks of the products and the companies were warned that any similar marketing in the future would lead to legal action.[17]

The code in industrialized countries

With the passage of the international code and the growing scientific evidence to show that even in industrialized countries, bottle-fed babies are likely to be less healthy than breast-fed babies, NGOs in those countries had a new focus for campaigning. There was an opportunity to bring the issue home by pressing for adoption of the code and changes in hospital practices in the industrialized countries.

In the United States, Public Advocates, a law firm specializing in social issues, drew up a comprehensive analysis of the infant-feeding trends, health hazards of formula use, factors that discourage breast feeding and industry practices throughout the country.[18] The dossier they prepared formed the basis of an administrative petition from thirteen women's, health and public interest groups, including ICCR and INFACT, which called on the United States' government to develop a national strategy for the promotion of breast feeding, to introduce changes in hospital practices and to introduce mandatory changes on labels of infant formula sold in the United States.

The United States' government, having just voted against the international code, was reluctant to open up an investigation of infant-formula misuse within its boundaries. However, the petition did spark off a set of congressional hearings in June 1981, and it received press coverage across the United States.[19] Although not all of the requests in the petition were met, the US Food and Drug Administration (FDA) did introduce, in December 1982, a compulsory warning on product labels to state that before using the product, mothers should seek the advice of a health professional. The FDA also took note that there was a need for a graphic symbol on labels to distinguish between powdered formula, liquid concentrate formula (which requires dilution with water), and liquid, ready-to-feed formula. However, the FDA suggested that this could be left up to the industry to use the graphic symbols voluntarily.[20]

Recognizing that national implementation of the code is unlikely in the United States, pressure groups such as INFACT began to concentrate on local implementation of the code. The strategy looks like gaining its first success in Washington, DC where, in April 1983, the City Council began consideration of an Infant Feeding Rights Act. The act incorporates recommendations on breast feeding from the American Academy of Pediatrics and was prepared with the assistance of INFACT. If it is approved, it will require Washington hospitals to adopt policies which protect a mother's right to:

- breast feed as soon after delivery as possible;
- practise on-demand nursing without supplemental bottle feeding;
- choose 24 hours rooming-in (have her baby in the room with her);
- avoid lactation suppressants unless they are administered with the mother's informed consent.

The act would also prevent the distribution of free samples to mothers who have not given informed consent to formula feeding.[21]

A major method of bringing home the need for the code in industrialized countries has been for NGOs to undertake monitoring surveys of promotional practices. In the United Kingdom, War on Want, together with Oxfam, BMAC and 3W1 conducted a country-wide survey in late 1981. In a six-week period the survey found nearly 1,700 examples of practices banned by the international code, and the groups used the publication of the results as a means of putting pressure on both the companies and the British government.[22]

The response of the companies and the government was far from encouraging. During 1982, the British Food Manufacturers' Federation (FMF), had talks with the Ministry of Agriculture, Fisheries and Food and the Department of Health and Social Security (DHSS), in which they began to formulate a code of practice for the United Kingdom. The NGOs and several health-worker organizations repeatedly asked the DHSS to open up those talks to include all groups interested in the issue. The DHSS refused, and a few days before Christmas in 1982 issued a draft code for 'consultation'.

The 'consultation' process consisted of an eight-week period during which different organizations were invited to submit written comments on the proposed code which the government claimed would 'fulfil its undertaking to implement the aim and principles of the international code'.[23] War on Want immediately issued a detailed commentary on the draft code, terming it 'totally inadequate'. The War on Want document found twenty-seven substantive differences between the two codes, with the most obvious being that the FMF code would allow manufacturers to continue to advertise and promote their products within the health care system, and in magazines and other publications aimed directly at parents.[24]

Other organizations such as the Health Education Council, the Health Visitors' Association, many Community Health Councils, the National Childbirth Trust, Oxfam, BMAC and individual health workers also condemned the FMF code.[25] Despite this strong protest, on 2 August 1983, the DHSS brought into effect the FMF code. Its final version still had twenty-four substantive discrepancies with the WHO code.[26] The DHSS promised that it would set up a Monitoring Committee to ensure the code was put into practice. However, it was not until late 1984 that the composition of the committee was announced and it did not hold its first meeting until March 1985. No public report has been issued of that meeting. In the meantime, War on Want launched a nation-wide campaign to have the FMF code withdrawn, a campaign which has been supported by MPs from all parties, many health organizations, and thousands of individual health workers.

At the same time as the FMF was working on a weak code for the United Kingdom, its counterpart organization within the EEC—the Association of Dietetic Foods Industries of the EEC (IDACE)—followed the same pattern. In October 1981, the European Parliament approved a strongly-worded resolution calling for a Community-wide directive to enact the international code. The European Commission responded with a proposal that a directive could be prepared to deal with labelling and quality control, but that a voluntary code would be the best way to deal with the marketing. IDACE obliged with a draft code virtually identical to the one the FMF prepared for the United Kingdom. The European Commission seemed satisfied with the IDACE code. Development agencies, health workers and consumer groups in Europe were less than satisfied. War on Want prepared a critique of the IDACE code which it circulated to members of the European Parliament (MEPs), NGOs in Europe, the media and to relevant Commission members.[27] The Bureau Européen des Unions des Consommateurs (BEUC), which had earlier cooperated with War on Want in a monitoring study of promotional practices in Europe, submitted a detailed critique of the IDACE code via the Consumer Consultative Committee of the EEC, of which it was a member.[28]

The controversy led the European Parliament to debate the issue again in April 1983. The Parliament once more overwhelmingly passed a strong resolution calling on the Commission to issue a directive on the application of the international code. It also called on the Commission to 'adopt a more constructive attitude and to be more heedful of the relevant resolutions adopted by Parliament'.[29] The Commission now appears to have distanced itself from the IDACE code. It issued a draft directive on labelling and quality control for discussion in late 1983, but made no mention of the IDACE code, which in the meantime has been revised by the industry. The revision eliminates any mention of links with the EEC action, and it appears that the EEC will simply stick with the limited directive, and let the industry issue a voluntary code if it wishes to do so, but will not give any such code a 'stamp of approval'.

An evaluation of national efforts?

What was clearly emerging as attempts were made to bring the code into effect was a strong pattern of resistance by the industry, particularly in the most valuable markets. Europe and North America were two industrialized areas where the industry did not want to see strong codes. In the developing world, South East Asia, India, Kenya, Nigeria, South Africa, Brazil and Mexico were the areas where codes were most hotly contested.

This resistance was coupled with the slowness of bureaucracies in many countries as ways of implementing the code were being studied. Governments found it easy to vote for a code at an international assembly, but considerably more difficult to put the code into practice. Decisions had to be made about whether to legislate or opt for voluntary agreements; whether to modify the code or adopt it completely unchanged; which ministries or departments should be included in the discussions.

Twelve months after the code was approved internationally, only one country—Trinidad—was able to announce that the code was in effect. Trinidad adopted the code, in its entirety, as a voluntary measure, and began work on the development of legislation to cover the requirements on labelling.

WHO, as a result of the resolution passed in May 1980, was mandated to present a report to the WHA every two years, 'on the steps taken by WHO to promote breastfeeding and to improve infant and young child feeding, together with *an evaluation* of the effect of all measures taken by WHO and its Member States' (emphasis added).[30] The twenty-eight page report prepared for the 1982 WHA contained no such evaluation, but did conclude that there was 'evidence that many countries have taken relevant action'.[31] The information provided for most of the seventy countries referred to in the report noted that a committee had been established to study how best to proceed.

NGOs were appalled at the report. It was seen as very bland, non-evaluative and lacking any spur to action. The whole tenor of the report was that these things take time, and that countries were, by and large, doing a good job in implementing the code. The lack of evaluation meant, for example, that Malaysia and Singapore were reported as having codes in effect. What went unreported was that those codes had been written several years previously, were heavily influenced by the infant-food industry, and bore little or no resemblance to the international code. A casual reader of the report would have been left with the impression that a greal deal had happened in the twelve months since the code had been approved.

The 1982 WHA was not expected to spend much time on the infant-feeding issue. After the previous year's controversy, WHO staff anticipated little more than a brief discussion of the report, a murmur of approval, and work would continue. The review promised in 1983 was supposed to be the next major discussion on the code. Another factor which minimized the likelihood of much discussion on infant feeding was the agenda item on the WHO 'Action Programme on Essential Drugs and Vaccines'. Journalists, delegates, the pharmaceutical industry, WHO staff and NGOs were expecting this to be the 'hot' issue. Health Action International (HAI), which was formed in May 1981 as a network of NGOs similar to IBFAN, mounted a major pressure campaign calling for a code to control the inappropriate marketing, promotion and distribution of pharmaceuticals. The precedent had been established with the infant-feeding code, and earlier WHA resolutions on drugs had mentioned the possibility of a code on drugs.[32]

Infant feeding became an 'old' issue. Delegates to the WHA listened sympathetically to the members of the IBFAN team who talked with them, but saw little they could do, or indeed, little that needed to be done at the assembly. A few, like India, included references to the code in speeches, noting that 'more rapid action is urgently required. Progress appears to be far from satisfactory ... Meanwhile, commercial firms continue vigorous campaigns for marketing substitute foods'.[33]

The issue would not have received much attention if it had not been for one of the delegates from Qatar, A. Mawlawi, who was attending the WHA for the first time. He felt it was an important issue and doubted if a few speeches would have much impact. At the end of the first week of the assembly, he suggested to members of the IBFAN team that a strong resolution would be more effective, and offered to propose such a resolution.

Although a newcomer to the assembly, he was no stranger to politics and negotiation. The draft resolution *was* strong; it contained a clear message that few member states had actually done what they said they would in voting for the code, and the implication that WHO was not doing enough to support efforts to implement the code. At the start of the second week of the WHA, he quietly circulated the draft resolution to delegations he felt would support it. By the end of that day, twelve countries had given support and he submitted it to the assembly secretariat for open distribution. It was the first that WHO staff, and most of the industrialized countries, knew of the resolution. It sparked off confusion and anger, for the resolution had not been prepared in consultation with WHO's staff, unlike most resolutions at the WHA. It was outside the normal system of control and came as a complete surprise. When WHO staff suggested a meeting with Mawlawi to offer some modifications to the resolution, he refused, claiming that the resolution contained precisely what he wanted to say, and that there was no need to discuss it further.

The United States, particularly sensitive on this issue, approached WHO staff to find out what was going on, and discovered that the traditional sources of information were not of any use in this case. WHO staff were powerless to mediate, so the United States, with several amendments it wanted to see made to the resolution, arranged a meeting with Mawlawi. The United States also invited Dr J. Cohen, special assistant to WHO director-general Mahler, to the meeting. Mawlawi requested that Cohen leave, as it was to be a meeting between delegates from two sovereign nations and there was no need for a representative from WHO. It was a further shock to WHO and an unsettling manoeuvre for the United States. Embarrassed by the rebuke, and nervous about its past reputation on the infant-feeding issue, the United States found it difficult to negotiate more than a single word change in the resolution. Only the deletion of a reference to the intention to *regulate* marketing could be obtained in return for American support for the resolution.

The NGOs were under heavy pressure as well. WHO staff, plus some industrialized countries' delegates, recognizing that IBFAN was supporting the resolution and lobbying additional support, urged IBFAN to persuade Mawlawi to modify the resolution. IBFAN refused, pointing out that a strong resolution was precisely what was needed to keep this issue a priority. WHO staff were convinced that the resolution would divide the assembly; it would spoil the near consensus on the code; and, besides, it was the wrong year to highlight the code—that would come in 1983.

In the end, the resolution was adopted unanimously. In addition to calling for member states to give renewed attention to implementing the code, it requested the WHO director-general to 'design and coordinate a comprehensive programme of action to support Member States in their efforts to implement and monitor the Code and its effectiveness', and to provide support and guidance 'to ensure the measures they adopt are consistent with the letter and spirit of the International Code'.[34]

The last point was particularly important as a new strategy for dealing with the code was introduced by Nestlé during 1982.

Notes

1. IBFAN, press statement, 'Baby food action groups celebrate code victory, intensify pressure', Geneva, 27 May 1981.
2. Ibid.
3. I. Guest, 'Milk and water reaction to baby food code', *Guardian*, 11 February 1981, p. 8.
4. *IBFAN News*, April 1983, p. 1.
5. *INFACT Update*, May 1981, p. 1.
6. An Early Day Motion is a British parliamentary device which calls upon the government of the day to undertake some specific action or give a commitment on a particular issue. It is a way in which topics that are not part of the scheduled parliamentary business can be introduced. If sufficient MPs support an EDM, it can lead to either an adjournment debate (10–15 minutes' debate at the end of a day's business) on the issue, or, occasionally, to a full debate. The more MPs supporting such a motion, particularly if the support is from all parties, the more likely it is that the government will act.
7. R. Deitch, 'Marketing of breast-milk substitutes in developing countries', *Lancet*, 7 March 1981.
8. Taped interview with Edward Baer, 27 September 1980.
9. Tape of IBFAN press conference, Geneva, 15 May 1981.
10. IBFAN, *Report of the First Congress*, Geneva, June 1981.
11. UNICEF internal memo on 'Meeting of the Coordinating Council of IBFAN in London, 23–27 Feb. 1982', New York, 22 March 1982. (In 1984, the same observer commented: 'IBFAN is in the process of transformation from a loosely organized but narrowly focused coalition to a more structured network encompassing a wide range of goals and objectives. In doing so, its members are dealing with issues of hierarchy, accountability and a more secure financial base—questions which rarely arose when the overriding priority was to last from one World Health Assembly to the next. The transition is not over and the results of the Mexico decisions, i.e. to formalise the network and appoint a General Administrator, are not yet clear. The essential point, however, is that the IBFAN Coordinating Council is aware of the need to reassess its present structure and adjust to new conditions. If the decisions of Mexico are not the right ones they can be reconsidered, meanwhile the size and capacity of the network will continue to grow. The vitality, dedication and effectiveness of IBFAN—both as a collection of outstanding individuals and as an international network—remain unchanged and as vibrant as ever.'—UNICEF internal memo, 'Meeting of Coordinating Council of the International Baby Foods Action Network (IBFAN), 29 Jan.–1 Feb. 1984', 21 March 1984.)

12. The 'Action Pack'—*Breast is Best: From Policy to Practice*—is a comprehensive guide to the issue, with a particular focus on the international code and ideas for implementing it nationally. The pack also contains additional resource material—photos, cartoons, posters, illustrated articles—and has been widely used by NGOs, health workers and the media for getting the issue across. The pack was published in mid-1982 in three language editions, English, French and Spanish, and by mid-1983 was being distributed and used in more than 100 countries. A modified Arabic version of the pack was produced in 1984.
13. *Business India*, 'The dirty sales tricks of baby food giants', 18 January 1981; and *IBFAN News*, February 1981, April 1981.
14. See R. K. Rangan, 'Austere paediatric meet', *Times of India*, 23 June 1983, p. 1; and IBFAN, *Infant Formula Promotion 1980*, Geneva and Minneapolis, January 1981. (It is worth noting that the American Academy of Pediatrics (AAP) was much more willing to accept funding from the infant-formula industry. At the end of 1981, the AAP agreed a two-year contract with Abbott/Ross Laboratories valued at $500,000 per year, which would provide partial support for the publication, *Pediatrics in Review*. In addition, Abbott/Ross would be allowed to commission up to four special educational items for AAP members. The main emphasis of the support was to ensure that all third-year residents in paediatrics would receive the materials. *California Pediatrician* commented that this would have the 'advantage of introducing the Academy to paediatricians while still in training'. It could be added that it would also have the advantage of introducing Abbott/Ross to paediatricians. The AAP felt that 'the necessary safeguards appear to be in place' with regard to the industry funding. For further details, see *California Pediatrician*, 'Education Grant', winter 1981, p. 9.)
15. See J. Worrall, 'Baby food code welcomed', *Financial Times*, 27 April 1981; J. Worrall, 'Baby milk curbed in Kenya', *Sunday Times*, 26 April 1981; *Standard* (Kenya), 'Vital ethics of infant care', 18 April 1981; D. K. Munyakho, 'Kenya goes easy on baby milk code', *Daily Nation* (Kenya), 25 January 1983.
16. *IBFAN News*, May 1983, p. 1. (CCF decided on the new brand name, Dutch Lady, and planned to have products with new labels on the shelves by April 1984.)
17. *Daily Times* (Nigeria), 'Two brands of baby food banned from circulation', 13 November 1981; *Daily Sketch* (Nigeria), 'Baby food banned', 13 November 1981; and J. Clark, 'Custard powder in Nigeria', *New Internationalist*, April 1982, p. 21.
18. L. Salisbury and A. G. Blackwell, *Administrative Petition to the United States Food and Drug Administration and Department of Health and Human Services*, San Francisco, Public Advocates Inc., 1981.
19. See, for example, A. Bonner, 'Improper use of baby formula not only an overseas problem', *Washington Post*, 19 June 1981; M. E. Schoonmaker, 'Baby formula: A peril in the US?', *Sunday Record* (New Jersey), 28 June 1981.
20. *INFACT Update*, January 1983, p. 4.
21. *INFACT Update*, June 1983, p. 4. (A similar local law was also under discussion in New York in April 1983.)
22. War on Want, *Breast or Bottle? Factors Influencing the Choice of Infant Feeding in the UK*, London, 1982.
23. DHSS, press release, 'Code of practice on breast-milk substitutes', 21 December 1982. (The closed nature of the process whereby the British code was developed is indicated in a letter from Norman Fowler, Secretary of State for Social Services, to Patti Rundall, chairperson of BMAC, in February 1982. Fowler explained that discussions were being held with the FMF and that 'guidelines *agreed* at these talks will be issued for consultation and we will certainly include your organisation in this. It would perhaps be preferable to postpone consideration of a meeting until you have had time to consider it' (emphasis added). In August 1982, War on Want obtained a copy of the first

draft of the FMF code, dated May 1982. War on Want immediately wrote to the DHSS protesting about the document and included a short analysis highlighting the discrepancies between it and the international code. A month later, Dr E. L. Harris, Deputy Chief Medical Officer, wrote back to War on Want stating that the document was 'a preliminary draft only', implying that there would be substantial changes in it. In fact there were fourteen changes—most of them grammatical and syntactical, with only two of any substance. The final version of the British code had a further three changes; thus, the document written by the industry in May 1982 went through the entire 'consultation' process with little amendment. However, had it been an open process, with the industry, health workers and NGOs all able to present arguments and counter-arguments, as happened in the consultations over the international code, the FMF code would almost certainly have been dismissed.)

24. War on Want, *Commentary on and Suggestions for Improvement of 'Draft Code of Practice for the Marketing of Infant Formulae in the United Kingdom and Schedule for a Code Monitoring Committee'*, London, January 1983. (Another discrepancy concerned the question of gifts to health workers. The international code said 'no financial or material inducements to promote products within the scope of this Code should be offered by manufacturers or distributors to health workers or members of their families, nor should these be accepted . . .'. WHO and UNICEF, in an explanatory note prepared in 1982, took this to mean a ban on *any* gifts, as it was virtually impossible to define what constituted an 'inducement to promote'—although this was subsequently modified by UNICEF in its discussions with Nestlé in 1984 (see Chapter 9 for details). The FMF code repeated the first clause of the provision, but omitted the second. In its place was a sentence allowing for 'articles of general utility . . . provided they are inexpensive and relevant to the practice of medicine and general health care'. There is a possibility that this passage in the FMF code might fall foul of the UK Prevention of Corruption Acts, 1906 and 1916, which prohibit health workers from 'soliciting or receiving any gifts or consideration of any kind from contractors or their agents, or from any organisations or individuals with whom they are brought into contact through their official duties as an inducement or reward for: (1) Doing/refraining from doing anything in their official capacity; (2) Showing favour or disfavour to any person in their official capacity'. One health authority in the United Kingdom notes that 'casual gifts', for example at Christmas, may not be connected with the performance of duties 'so as to constitute a criminal offence under the Act but nevertheless such offers should be politely but firmly declined'—see Ealing Health Authority, memorandum circulated 11 November 1982, to be included in the Authority's Standing Orders and Standing Financial Instructions. This raises important ethical and possibly legal questions.)

25. The Health Education Council said the FMF code 'is inadequate for our circumstances . . . We believe that a major reason for the inadequacy is the very limited consultation process used . . . The development of the draft code mainly by industry has resulted in a code that too often places commercial interests before health concerns . . . The WHO Code is contravened in several places', *Comments from the Health Education Council on the United Kingdom's Response to the World Health Organization's Code of Marketing of Breast-milk Substitutes*. The Pharmaceutical Society noted that 'health authorities and clinics should not receive samples of breast-milk substitutes or promote their use in any way. It is also contrary to the spirit of the code that persons engaged in marketing of infant formulae should be able to take part in the education of pregnant women or nursing mothers', 'Breast milk substitutes', *The Pharmaceutical Journal*, 12 February 1983. The National Childbirth Trust said, 'The FMF code as it stands needs substantial and drastic alteration to make it satisfactory. At present it is clear that it comes nowhere near to fulfilling the terms of resolution WHA 34.22, and we urge the DHSS and MAFF to take steps to implement this resolution fully and as soon as possible', *Commentary on*

the Code, Breastfeeding Promotion Group of the National Childbirth Trust, London, 10 February 1983. The Health Visitors' Association expressed 'overwhelming disappointment' over the code, noting that 'in spite of the wide consultation it seems as if none of the points raised were taken into account', 'HVA slams voluntary code on baby milk', *Nursing Times*, 10 August 1983. The British Medical Association said it was 'concerned that the effect of the British code, far from promoting breast feeding, as is the avowed intent of the government, will be the opposite', 'Baby milk code fails to calm critics', *New Scientist*, 11 August 1983. See also J. Nicholls, 'Pressure groups still urge tighter controls on baby food promotion', *Medical News*, 13 January 1983; D. Hencke, 'Britain "intends to break baby food health code"', *Guardian*, 15 January 1983; 'Substitute for breast milk escapes curbs', *New Scientist*, 20 January 1983; P. Chorlton, 'Sour reception for new code on powdered milk marketing', *Guardian*, 2 August 1983.

26. FMF, *Code of Practice for the Marketing of Infant Formulae in the United Kingdom and Schedule for a Code Monitoring Committee*, London, July 1983. (The FMF code bears a remarkable resemblance to voluntary codes that have been introduced in the Republic of Ireland (December 1982) and Australia (May 1983). Governments in both countries, as well as the United Kingdom, were reluctant to accept an unmodified international code because they claimed national circumstances were different. However, they appear to have no qualms about accepting a 'universal' model code prepared by the industry.)
27. War on Want, *Commentary on and Suggestions for Improvement of 'Code of Practice for the EEC for the Marketing of Breast Milk Substitutes'*, London, February 1983.
28. Commission of the European Community, *Preliminary CCC Opinion on the Implementation of the International Code of Marketing of Breast-milk Substitutes* (Doc. CCC/32/83 final), Brussels, March 1983. (The monitoring study referred to is available from War on Want, *The International Code of Marketing of Breast-milk Substitutes: Is It Being Followed in Europe?*, London, October 1982.)
29. European Parliament, *Minutes of Proceedings of the Sitting of Monday, 11 April 1983* (Doc. PE 83.818), Strasbourg, April 1983.
30. WHO, *Resolution WHA 33.32 of the 33rd WHA*, Geneva, May 1980.
31. WHO, *Infant and Young Child Feeding: Progress Report by the Director-General* (Doc. A35/8), Geneva, May 1982, p. 14.
32. WHO, Resolutions WHA 31.32 and WHA 32.41 in 1978 and 1979 laid the basis for the Action Programme.
33. Dr S. Sidhu, at the 35th WHA, May 1982, reported in *Baby Bottle Disease: A Call for Action*, Minneapolis, INFACT, 1982.
34. WHO, *Resolution WHA 35.26 of the 35th WHA*, Geneva, May 1982. See also D. Mackenzie, 'From the mouths of children', *New Scientist*, 17 June 1982. (An internal War on Want memo notes that the 1982 WHA changed the attitude of NGOs towards WHO considerably, beginning with a meeting which IBFAN representatives had with WHO staff during the first few days of the WHA. Dr D. Tejada, assistant director-general, told the NGOs that he would welcome closer contact with the industry, 'inviting them into the chicken coop so that WHO could control their activity'. The NGOs were appalled at the naïvety and pointed out that if that was the secretariat's attitude, it would receive little support from NGOs in future. The secretariat responded that its constituency was the member states, and that NGOs, while useful, were unimportant compared to the opinions of member states. The memo goes on to describe the various events which occurred over the resolution and concludes, 'The exercise also communicated a message to WHO staff—NGOs could not only deliver constituencies of support at the national level, but also deliver constituencies of support amongst Member States at the Assembly'.)

9. The Nestlé approach

> Basically, what affects people are matters of the heart. Whether or not Nestlé abides by specific articles of the Code is important to us, but what is important to people generally is what happens to babies.
>
> Jessica Govea, former organizer with the United Farm Workers and consultant to INFACT[1]

Two events occurred on 16 March 1982 in different parts of the world which are of significance to the infant-feeding campaign. The first incident, in Manila, was virtually unnoticed. The second, in Washington, captured headlines around the world.

Miranda de Campos was unaware of the international code when she entered Pasay General Hospital in Manila in January 1982 to have her third baby. Miranda breast fed her first two children, but with her third, Juanita, she bottle fed because when she left the hospital she was given a free sample of Nestlé's Lactogen formula and a glossy leaflet promoting its use. On 16 March, Miranda brought Juanita back into hospital. The child was severely malnourished, coupled with diarrhoea and vomiting. Nurses said the baby's feeding bottle was encrusted with dirt. Juanita went from the feeding bottle to another bottle, this one containing an intravenous solution to keep her alive.[2]

Meanwhile, in Washington, Rafael Pagan, president of the Nestlé Coordination Centre for Nutrition, unveiled to a large press conference 'comprehensive policy guidelines to be applied in countries which have not yet taken measures to implement the WHO International Code of Marketing of Breast-milk Substitutes'. The title of the press release given to journalists was carefully chosen: 'Nestlé completes WHO code implementation process'.[3]

The message Nestlé wanted to convey was that the controversy was over; the boycott could be called off; the company had changed its practices and would abide by the international code. For the first time in the ten-year campaign, Nestlé got its public relations approach just about right. The story received massive press coverage in the United States and elsewhere; at least 115 newspapers in the United States carried the story and countless others in different countries. The basic sense of the press reports was that Nestlé would abide by the WHO code, although INFACT noted that 90 per cent of the reports also contained the message that the boycott would continue.[4]

Together with the press release, journalists received an impressive thirty-two page document containing Nestlé's instructions to its staff about the code.[5] The essential element of the document was the reproduction of the international code, paragraph by paragraph, alongside 'itemised implementation instructions' for each provision. Pagan termed the approach by Nestlé 'a quantum jump on the whole issue, the whole infant formula controversy ... I can assure you Nestlé is implementing a very stringent code of checks and balances to make sure it works'. Leah Margulies of ICCR was less optimistic: 'My experience in the past with Nestlé is that they've made promises that haven't been kept'.[6]

Despite the fanfare, however, Nestlé's instructions came under critical scrutiny and were found to be insufficient. *The Lancet*, for example, said the instructions 'raise some doubts about the purity of Nestlé's intentions . . . Analysis of Nestlé's guidelines discloses ways in which they might be used to circumvent various parts of the International Code, such as the Code's stipulations on free samples to health workers and free supplies to institutions'.[7]

According to INFACT, the Nestlé instructions were a flawed interpretation of the international code, which attempted to:

- Narrow the range and timespan of products covered by the Code;
- Deny Nestlé's responsibility to provide detailed information of the hazards of formula misuse and the benefits of breast feeding, as required by the WHO Code;
- Legitimize practices outlawed by the WHO Code, including generic infant formula advertising, personal gifts to health workers and free formula for personal use by doctors;
- Routinize the provision of free formula supplies to hospitals while disclaiming industry responsibility for starting the practice; and
- Introduce ambiguity on the role of company marketing personnel, label content and Code implementation.[8]

A group of independent legal experts, who were attending a WHO meeting on the development of legislation to assist in the implementation of the code in June 1982, expressed similar reservations about Nestlé's instructions, citing eight specific points which they felt were at variance with the international code.[9]

Delegates from several developing countries at the 1982 WHA also drew attention to the weaknesses in the Nestlé instructions. Dr A. Maruping of Lesotho said the Nestlé interpretation had 'a number of unfortunate loopholes'; Dr G. Ondaye from the Congo warned that without national legislation 'we may fall into the trap of the large multinational companies which are proposing various "interpretations" of the International Code to be applied in our countries but which may, in fact, take into account more the commercial interests than those of health'; Dr A. Tarutia from Papua New Guinea said, 'the Nestlé document is a serious distortion of the World Health Organization Code. Marketing practices prohibited by the Code are approved of and encouraged by the Nestlé document. My delegation is convinced no private sector has the right to amend the WHO/UNICEF Code'.[10]

UNICEF issued a rebuke to Nestlé over the instructions, claiming it had 'serious misgivings on Nestlé's interpretation of significant aspects of the Code and on the possible harmful effects of its instructions in the implementation of the true spirit and intent of the Code'.[11] UNICEF and WHO issued a joint set of notes on the code which set out what they saw as the spirit and intent of the various provisions.[12]

When the instructions were first issued, Nestlé was adamant that they were non-negotiable; they represented the final concession the company was prepared to make.[13] However, faced with such strong criticism from all circles, Nestlé was forced to revise its instructions and, in October 1982, issued a second set which incorporated changes in policy that more closely resembled the provisions of the international code.[14] Despite the changes, the new instructions still fell short of the code. UNICEF, in a memo to field directors, referred to 'notable differences . . . the most important is that Nestlé does not follow a universal policy as regards implementation of the Code'.[15]

A new watchdog?

If there was widespread concern over Nestle's policy, there was even more scepticism about the company's willingness actually to put the policy into practice in the field. In order to deal with this doubt, Nestlé also announced that it would establish an independent commission to audit the company's performance with respect to its instructions on the code. On 3 May 1982 (the same day the 1982 WHA began), Nestlé announced the composition of the new commission, with former US Secretary of State Edmund Muskie as chairman.[16]

Although established with the basic objective of examining 'complaints and allegations about Nestlé's marketing practices', none of the members of the Nestlé Infant Formula Audit Commission (NIFAC) had any direct expertise with the techniques of marketing consumer goods. A further worry over NIFAC was that it had been selected by Nestlé and would be totally financed by Nestlé, thus casting some doubt on its impartiality. Indeed, the first question journalists asked Muskie when he gave a press conference in Geneva in mid-May 1982 was 'are you partisan on this?' In a long, rambling and antagonistic response, Muskie emphasized again and again that the charter of the commission would ensure its independence, and that all the members of NIFAC were determined to demonstrate their independence. 'Your question assumes we have no personal integrity. I'm not about to sell my credibility . . . We come to this without any strings that Nestlé can pull to influence our judgement.'[17] In a subsequent question, he was asked whether Nestlé had therefore agreed in principle to abide by NIFAC's findings and change practices which were leading to violations of the code. Muskie dismissed the question as somewhat theoretical, implied that there was no such undertaking and said, 'We may conceivably remain in disagreement with the company'. However, a few minutes later, he did an interview for Swiss radio when he said, 'None of us have any doubt at all that we will be dealt with by

Nestlé as an independent Commission, and that they, whatever the damage to the company, will honour the Commission's recommendations'.[18]

The main worry for the NGOs, however, was that NIFAC would serve as a very effective deflection shield for Nestlé. Rather than having to deal with complaints directly, the company could now say that NIFAC was studying the situation. The complaints procedure was carefully worked out. Upon receipt of a complaint, if it was deemed to be valid by the chairman, a set of questions would be prepared and submitted with the complaint to Nestlé. The company then had sixty days to respond. Upon receipt of the Nestlé response, a 'preliminary determination' would be prepared and sent to the complainant, asking for further information and offering the opportunity to rebut any company rationale for a particular practice. The complainant would be given thirty days to respond. NIFAC would then prepare a 'final determination' (or if necessary, go back to Nestlé for further information), which would be communicated to Nestlé together with recommendations for corrective action if the complaint was found to be valid. Nestlé would then have thirty days to communicate to NIFAC if it accepted the findings and give a firm date as to when corrective action would be taken. A final report would then be prepared and sent to the complainant, and generally be made public by its reproduction with the next scheduled quarterly report of the commission.[19]

At the very least, five months would elapse from the time a complaint was reported to the production of a final report on what was being done about it. In practice, the time-scale is even longer. Of nearly sixty complaints submitted to the commission in November 1982, only twenty-one had been completely processed by May 1983. Because of the quasi-judicial nature of the commission, the complaints procedure involved a great deal of paper work. For NIFAC, this was no great problem, with adequate staff paid for by Nestlé. For small NGOs without surplus administrative staff, it was more of a hardship.

The commission has taken a strong line with Nestlé on occasion. It has dismissed many of the usual arguments put forward by Nestlé: it's an old poster, we don't produce it any more; the labels are in the process of being changed; the national code in that country allows us to produce this type of material. NIFAC, while accepting that those arguments may be factually correct, does not accept them as justification for the company's behaviour. It has asked for specific dates on the introduction of new labels; has urged the company to stress with employees the need to remove any *old* materials still in circulation; and has said that approval of a particular practice by a national government 'does not excuse' a violation 'because Nestlé has agreed to comply with the stricter of a local code and the WHO code'.[20] However, NIFAC lacks any enforcement mechanism, and if the company chooses to do so, it can simply ignore the advice of the commission.

INFACT notes that while the commission may criticize from time to time, it 'does so softly . . . bending over backwards to give the benefit of any doubt to Nestlé'. In examining NIFAC's *Fourth Quarterly Report*, INFACT found it to be a curiously contradicatory document. Its *theme* is similar to that of NIFAC's other three reports: 'The Commission maintains its belief that Nestlé is committed to implementing the WHO Code'. Yet its *content* reveals that Nestlé has not implemented the Code and that, if present trends continue, it will not do so in the near future.[21]

That contradiction is the main bonus to Nestlé of the commission. Although NIFAC has agreed from time to time that Nestlé has some blemishes, it has done a great deal to polish the company's tarnished image. Muskie said in 1983, 'Nestlé's record is not perfect, but it is honouring, and attempting to honour, its publicly stated commitments'.[22]

The whole exercise has provided a new marketing gimmick for Nestlé. It can claim, and has done so, to be the first and *only* company to agree publicly to follow the international code, and to subject itself to a relatively public audit of its performance. It will never receive a clean bill of health from WHO—WHO does not comment publicly on the activities of individual companies—but Nestlé can imply as much, simply by pointing out that WHO has not criticized the company for its recent moves. It is the same marketing technique used to sell any consumer good where product differentiation is largely the result of a creative copywriter in the advertising department. In this case, the 'difference' was not so much to achieve better sales figures for infant formula—Nestlé has even accepted the possibility that strict adherence to the code might cost them sales in some countries.[23] The 'difference', however, was to take the pressure of the boycott off Nestlé. Carl Angst, executive vice-president of Nestlé, explained:

> Once you get to the point where a company has undertaken something which is possibly unique—I think it's almost unique in the free enterprise system—has gone out of its way to adopt policies, to change marketing procedures, then to show tremendous progress—I maintain, a tremendous progress—in achieving the objective, efforts which are being recognised today by a large number of churches, recognised by WHO, recognised by UNICEF, and recognised by other international institutions, once you are at this point, then, of course, I scratch my head and wonder, why do they continue amusing themselves with a boycott?[24]

If there was any doubt left after Angst's comments as to why Nestlé set up NIFAC, it was dispelled by the even more forthright answer another Nestlé official gave when NGOs asked why NIFAC was so heavily oriented towards the United States. He said, 'that's where the problem is—the boycott and the bad press'.[25]

Maintaining the boycott

Nestlé undoubtedly had taken major strides forward in its approach to the infant-feeding issue. When the boycott continued, the company began to suggest that it was being made a martyr. It wondered why the campaign was still so strongly focused on Nestlé, while other companies, who were not prepared to make similar concessions, seemed to get away with unacceptable marketing practices.

The reasons for continuing the boycott were essentially the same as the reasons for its initiation: Nestlé had the biggest market share and its practices affected more people. Despite the changes, NGOs found the new policy 'an inaccurate and inadequate interpretation' of the code; the practices still violating the provisions of the code; and were less than confident that without the external pressure of the boycott there would be any long-term institutional agreement that the code had to be followed, or systematic monitoring to see that it was. ICCR saw the continuation of the boycott as a way of putting pressure on the other companies as well:

> It's not just Nestlé we're pressuring . . . The extent to which Nestlé practices are brought into conformity with the Code becomes the new maximum standard for the entire industry and probably for most governments as well. Thus, we're not just focusing on Nestlé but we're setting the level of acceptable marketing practices for an entire industry for generations.[26]

That such a claim could be made was reinforced by the failure of the 1983 WHA to recommend any changes in the code, based on efforts to get it implemented. The WHO director-general's report, first discussed by the WHO executive board in January 1983 and forwarded to the 1983 WHA, noted that no member state had proposed any revision of the code, either in form or content and suggested therefore that 'it would be premature, at this time, to propose any revision'.[27] WHA delegates went along with that suggestion.

The reality of the situation was, therefore, that renegotiation of the code at an international level was extremely remote. For all intents and purposes, the document approved in May 1981 was a final document, unlikely ever to be modified. Individual governments might conceivably strengthen parts of the code in their own national settings (as India and Sri Lanka intend to do), but most would accept the code verbatim, or adopt a weaker version. (Table 6 gives a picture of the situation that existed in 134 countries in February 1985 with regard to the code implementation.) A few countries might introduce legislation; most would opt for a voluntary measure of some sort, such as occurred in New Zealand, where the government persuaded manufacturers to accept the international code in its entirety.[28]

Thus, the degree to which the manufacturers were prepared to accept the provisions of the code was likely to be the determining factor in how effective the code would be as a means of achieving changes in marketing practices. The company prepared to make the most concessions, in this case, Nestlé, would set the mark for the others to follow. It was therefore very important that Nestlé was pushed as far as possible towards total acceptance of the international code in both its policies and practices.

In the United States, INFACT began in early 1983 to concentrate the boycott on just one Nestlé product—Taster's Choice instant coffee—as a means of intensifying the pressure on the company. Coupled with strong, local campaign centres organized in the major cities, the Taster's Choice campaign brought quick results. In Chicago, for example, two independent supermarkets took Taster's Choice off the shelves, replacing it with a large poster explaining the campaign and the reasons for boycotting Nestlé.[29]

Other countries planned to use Nescafé as the basic target. In May 1983, representatives from boycott groups in the United States, Canada, the United Kingdom, Sweden and Germany met together to plan coordinated activities. It was the first such meeting and was prefaced by a successful international petition drive. During the first few months of 1983, groups in these and other countries collected signatures on a petition calling on Nestlé to abide by the international code in full. The petitions were delivered by an international delegation to Nestlé's annual general meeting in Lausanne, Switzerland in May 1983. In all, 112,000 signatures were collected from people in thirty-eight countries. The petitions were wheeled up to the door of the meeting hall in a baby's pram and handed over to astonished Nestlé employees in a short presentation that was filmed by Swiss television. The petition delivery received widespread press coverage and helped to remind people around the world that the boycott was continuing.[30]

It was decided to continue the international petition drive and to make further deliveries to Nestlé throughout the year. The groups also decided to present a strong united front in the face of attempts by Nestlé to negotiate with individual groups for an end to the boycott. A joint letter was drafted and sent to Nestlé informing the company of the united position and of the commitment to intensify and expand the boycott internationally, and calling on the company to open up serious negotiations with the International Nestlé Boycott Committee (INBC), the group which was set up in 1979 to represent the various boycotting organizations in any negotiations with the company.[31]

Blunders

The fourteen-month period from the time Nestlé announced its instructions until the boycott groups delivered the international petition was one in which the balance swung back and forth between the two parties as to where control over the popular campaign lay. The Nestlé instructions, establishment of NIFAC, the revised instructions and the first reports from NIFAC all tended to swing the balance towards Nestlé. As Rafael Pagan, president of the Nestlé Coordination Centre for Nutrition, said, 'The Commission holds the critics accountable and that has helped defuse the criticism'.[32]

As the NGOs began to revitalize the campaign in the early part of 1983, however, Nestlé became involved in three public relations blunders which helped to cast doubt on the company's intentions. The first incident involved a

Table 6 Progress in implementation of the code

Afghanistan	5	Ghana	5	Papua New Guinea	8
Algeria	X	Grenada	5	Paraguay	5
Antigua	6	Guam	5	Peru	10
Argentina	10	Guatemala	10	Philippines	7(10)
Australia	7	Guinea	8	Portugal	8/9
Austria	7/4	Guyana	6	Qatar	6(10)
Bahrain	5	Haiti	8/5	Romania	X
Bangladesh	8	Honduras	5	Rwanda	6(10)
Barbados	5	Hong Kong	8/7	Samoa	6
Belgium	8	Hungary	X	Sao Tome & Principe	8
Belize	5	India	(10)	Saudi Arabia	8/5
Benin	8(10)	Indonesia	6	Senegal	5
Botswana	8	Iran	X/6/5	Sierra Leone	6(10)
Brazil	4/8	Iraq	6(10)	Singapore	3
Brunei	7	Ireland	3	Spain	8/5
Bulgaria	X	Israel	6/8	Sri Lanka	10
Burkina Faso	5	Italy	5/8	St Kitts	6
Cameroon	6	Ivory Coast	7	St Lucia	6
Canada	7	Jamaica	6	St Vincent	5
Central African Rep.	6	Japan	8/7	Sudan	4
Chile	7	Jordan	6	Swaziland	6
China	X	Kenya	10	Sweden	9
Cook Islands	7	Korea, Rep. of	7	Switzerland	3
Colombia	8	Kuwait	5	Syria	9
Congo	6	Lesotho	6	Tanzania	5
Costa Rica	6(10)	Liberia	6	Thailand	9
Cuba	X	Libya	6/X	Togo	X
Cyprus	5	Malawi	5	Tonga	9
Czechoslovakia	X	Malaysia	3/4/7	Trinidad	9
Democratic Yemen	X	Maldives	1	Tunisia	8/6
Denmark	7/4	Mali	5	Turkey	4
Dijbouti	8	Mexico	6	Turks & Caicos Is.	5
Dominica	6	Mongolia	X	Uganda	6
Dominican Rep.	4	Montserrat	6	USSR	X
Ecuador	6(10)	Morocco	5	United Arab Em.	6/8(10)
Egypt	8	Mozambique	6/8	UK	3
El Salvador	7	Nepal	6	USA	7
Ethiopia	6/8	Netherlands	7	Vanatu	1
EEC	2	New Zealand	9	Venezuela	8
Fed. Rep. Germany	3	Nicaragua	8	Vietnam	X
Fiji	6	Nigeria	3	Yemen Arab Rep.	5
Finland	7/8	Norway	7	Yugoslavia	9
France	7/8	Oman	6(10)	Zaire	8
Gabon	7	Pakistan	6	Zambia	6/8
German Dem. Rep.	X	Panama	8	Zimbabwe	6(10)

This chart provides information on action taken in 134 countries and territories as well as the European Community. The best status—the international code in effect as law—has been given a score of 10. This score in brackets indicates that such legislation is intended, but has not yet been enacted. Reference to more than one number or letter indicates that either the country is moving from one stage to another or that there is implementation by segments.

Key:
0 — No action planned.
1 — No action planned, but situation closely watched.
2 — Voluntary code prepared by industry under discussion.
3 — Voluntary code prepared by industry in effect.
4 — A code under discussion incorporating some, not all, of the provisions of the international code.
5 — International code being studied by committee, working party, etc.
6 — International code recommended for implementation by working party; awaiting implementation.
7 — Voluntary national code or voluntary guidelines incorporating some, not all, the provisions of the international code in effect.
8 — Legislation, rules, regulations incorporating some, not all, the provisions of the international code in effect.
9 — International code in effect as a voluntary measure.
10 — International code in effect as law.
X —Government controls distribution/marketing.

Source: UNICEF, *Ideas Forum*, No. 19 1984, p. 18 and IOCU, *1985 Status of the Code by country* (as per 1 Feb. 1985), Penang, 1986.

symposium on ethical dilemmas which was sponsored by the National Institutes of Health (NIH) in the United States during January 1983. Included on the agenda was a discussion of the infant-feeding issue and Nestlé had initially agreed to take part. Ten days before the talk, the company withdrew and five days later sent a letter to the Secretary for Health and Human Services, with copies to two White House officials. The letter, written by John Mongoven, vice-president of the Nestlé Coordination Centre for Nutrition, set out why Nestlé decided not to participate. The basic reason was because the company felt the meeting was 'stacked'. Patricia Young, of the United Presbyterian Church and a board member of ICCR, was to give the history and background to the controversy, while Edward Baer of ICCR was to be one of the panelists.

In his letter, Mongoven described Baer as a 'leading political activist . . . Mr Baer's credentials include no scientific or professional expertise in the field, in our opinion, and represent only a long history of anti-corporate activism'. Young was described as a 'non-scientist whose only significant experience is as a board member of INFACT [factually incorrect], an anti-industry organization which has historically misrepresented the facts in this issue in general and our company's policies in particular'. He included a copy of an article describing Young as a 'scourge of the multinationals'. NIH staff pointed out that scientific qualifications were not necessary for the talk; it was meant to be about the ethical conflicts and lessons to be learned from the ten-year controversy. Nestlé did not ask that the meeting be cancelled. None the less, three days before it was to be held, the segment on the infant-feeding issue was taken off the programme.[33]

The second incident involved an offer by Nestlé to invest $1 million in a fundraising campaign for the British-based Save the Children Fund (SCF). The campaign would have been run in the United States and, according to an SCF official, would have included television advertising along the lines, 'Nestlé is helping to save the children: this time is paid for by Nestlé'. SCF declined the offer 'as association with Nestlé would do no good to the fund' because of the controversy over the infant-feeding issue. According to the *Guardian*, 'earning the goodwill of a charity like SCF could have helped Nestlé in its attempt to counter an international boycott of its products'. Gerry Raffe, general manager for Nestlé in the United Kingdom, admitted talking with SCF officials, but said 'I have no personal recollection of discussing an advertising campaign'.[34]

The third incident did not appear to involve any direct action by Nestlé, but was probably the most damaging to the company's image. The actress, Liv Ullmann, was on a goodwill tour through parts of Asia for UNICEF in early 1983. One of her stops was in the Philippines where she was scheduled to appear on a television programme, *Ms Ellaneous*. On the day before the programme was due to be filmed, the scriptwriter for the programme phoned the UNICEF office in Manila and said that Ms Ullmann could not appear on the show because the main sponsor, Nestlé's subsidiary in the Philippines, Filipro, 'was afraid that she would promote breast feeding'. UNICEF agreed to cancel the appearance.

Later that afternoon, the scriptwriter phoned back to say that 'Filipro had written to them instructing them not to tape Ms Ullmann'; however, she could appear provided she did not talk about breast feeding. UNICEF decided that in the light of the views of the sponsor and the restrictions imposed, Liv Ullmann would not take part in the show. The following day, a production assistant on the programme phoned UNICEF to confirm the taping schedule for the afternoon, unaware that the appearance had been cancelled. When she heard what had happened, she told UNICEF that the scriptwriter had no authority to cancel the appearance and that Liv Ullmann was welcome no matter what she wanted to discuss. By this time, Ms Ullmann's schedule had been rearranged, and it was not possible for her to appear. UNICEF also declined the offer to tape the programme at an alternative time.[35]

Ms Ullmann did, however, give a press conference in Manila and said, 'I am shocked that I should be censored. This has not happened anywhere else in my travels for UNICEF. These infant formula companies have become so big they are able to control media. I promise you it will be known all over the world'.[36] It was. Press wire services picked up the story and it was soon appearing in newspapers around the world.

A spokesman for Filipro said 'the company had nothing to do with Ullmann's failure to appear on the programme'.[37] Nestlé's executive vice-president, Carl Angst, phrased it slightly differently:

> The show was cancelled and it was then said that it was cancelled under pressure by the local Nestlé company, which, I think, to the best of our knowledge, after very careful investigation, by the way, this is strictly not correct.[38]

Angst was not correct in all the details of the incident when he explained it to Washington journalists a month later. He claimed that the incident 'happened before her arrival', when, in fact, she was in the Philippines when it occurred. He also said:

> for me, the incident is closed. For UNICEF management, the incident is closed. And for Ms Ullmann, the incident is closed. She got the whole story much later after she left the Philippines and I think she declared herself entirely satisfied by the declarations made. I think by now, she regrets this whole thing as much as we do.[39]

Two months later in London, Liv Ullmann held a press briefing for British journalists on the work of UNICEF, and mentioned the incident in a way that did not suggest she was entirely satisfied'.

> That was the dumbest thing they could do . . . We could say all this at the press conference, that the TV programme had been cancelled because a sponsor of the programme belonged to a milk company . . . I now think that in the Philippines they will be very careful with censorship both in newspapers and on TV when it comes to these affairs; there has been a lot of censorship, especially on TV, when people have tried to promote breast feeding; they have literally been stopped and letters have been sent to TV stations offering bribes in exchange for stopping programmes advocating breast feeding.[40]

The incident had an unexpected result by winning a new advocate for breast feeding—Philippines President Ferdinand Marcos! Marcos ordered an investigation of the incident and said:

> This is, from my point of view, an abusive use of the dominant power of a 'multinational' over the national interest of the host country. We are a host country, yet they are trying to dominate us, impose on us matters which conflict with out national interest and with the human rights and interests of individual citizens.[41]

The three incidents together provided fresh understanding of the power and influence that a multinational company like Nestlé could wield, even though, in every case, Nestlé actually denied doing anything directly to change the course of events. It is widely recognized that in a complex negotiation involving a shift in the balance of power, no matter what the level, the hints, the innuendos, the implied suggestions, are the materials that help to shape the final outcome. A vague suggestion that a company might not be pleased to have a television programme which it sponsors deal with certain subjects is often all that is required to ensure an over-cautious form of self-censorship by the programme staff. A letter, which does no more than suggest political motivations of certain individuals appearing in a government-sponsored forum can be enough to have a politican begin to worry about future repercussions and cancel the activity. The techniques for introducing these suggestions are similar to the techniques used to sell consumer products. A research director for J. Walter Thompson noted, 'What advertising does is reinforce messages that began years and years ago, something that people intuitively do'.[42] When applied to the world of policy-makers, the suggestions reinforce the status quo and help to maintain existing power blocks.

Activities like the Nestlé boycott offered an opportunity to push against the status quo. Nestlé had changed, but, as the three incidents showed, the fundamental motivations of the company and its methods of using, albeit indirectly, the power it possessed had not changed. As that recognition was grasped by people in the boycott movement, the impetus to carry on the struggle became stronger, and the balance of control over the issues of the campaign began to shift back to the NGOs.

Anwar Fazal, president of IOCU, called the boycott the 'first major global victory of health over profit, a victory that involves the protection and promotion of a very natural superior product: breast milk'. But, he added, although Nestlé had made substantial progress, the boycott had to be carried on to its conclusion 'because you cannot stop these kinds of campaigns 90 per cent of the way'.[43]

There was another reason—the most important reason—for the boycott to continue. Perhaps NIFAC could wade through the quasi-legal determinations of whether Nestlé was abiding by Articles 4.2, 5.1 or 7.2 of the code, but its jurisdiction did not stretch to dealing with the effects of any infringements of the code. It would not have been much comfort for Juanita de Campos, or her mother, if some six months after it happened, NIFAC decided that giving the sample which led her to the rehydration bottle was a violation of one of the articles of the code. It was a violation of something much more basic than that.

The complaints procedure focused on abstracts; the boycott on people. And despite all the claims, people were still suffering. In January 1983, Mark Torres was born in a Manila hospital. Mrs Torres, like Mrs de Campos, had breast fed her previous baby, and she, too, when she left the hospital, received a sample of Nestlé's infant formula. In February, Mark was back in hospital, suffering from the effects of bottle feeding.[44]

The answer to Carl Angst's question, 'Why do they continue amusing themselves with a boycott?', could be found in any one of several hundred paediatric wards throughout the Third World at the time he asked it. It was not an amusing answer.

Settling the boycott

Despite this, cracks were beginning to appear in the boycott movement by the middle of 1983. At grass-roots level in both Europe and North America, the boycott was getting stronger. But, particularly in the United States, where the boycott had originally depended on institutions for its strongest support, there were signs that some of them were beginning to waver. It was no accident that this was happening. It was part of a deliberate strategy by Nestlé to split the boycott movement open, as Rafael Pagan explained:

> Our primary goal is survival. Our secondary goal must be to separate the fanatic activist leaders—people who deny that free wealth-creating institutions have any legitimate role to play in helping the third world to develop—from the overwhelming majority of their followers—decent, concerned people who are willing to judge us on the basis of our openness and our usefulness. In particular, we must strip the activists of the moral authority they receive from their alliances with religious organizations.
>
> When I came to Nestlé last year, I realised that the company's highest priority was to establish a dialogue with those church leaders who were supporting, or who might support, the activists' boycott against us.[45]

That strategy led to the United Methodist Church in the United States in April 1983 that it would not formally endorse the boycott after a two-year investigation. It was left up to individual sectors and agencies of the church to decide for themselves whether to boycott. However, at the end of 1983, the United Methodist General Council of Ministries approved a report to go to the May 1984 General Conference which recommended that it would be inappropriate for any agencies of the church to take part in the boycott. The Church of the Brethren, a much smaller American church, also decided at the end of 1983 to withdraw from the boycott. Nestlé said of this:

> It is a simple fact that those who have made the greatest effort to root out the facts behind the controversy and confrontation are the very people who now say the time for controversy and confrontation is past. We find these churches' openness and honesty refreshing. We hope others do, too.[46]

Other groups were also reconsidering. The American Federation of Teachers voted to pull out of the boycott in January 1983; the National Order of Women Legislators decided not to join the boycott in September 1983; Georgetown University's student union decided to end its two-year participation in the boycott in October 1983.[47] Other groups and churches indicated to INFACT and the INBC that they would be assessing their participation in the boycott during the early part of 1984.

The INBC recognized that the trickle of defections could rapidly become a flood, and that, despite the increasing individual support for the boycott being generated at grass roots level in the United States and the growing support in Europe, the boycott campaign was in danger. Nestlé was making, and would continue to make, public relations gains over every organization that withdrew from or refused to join the boycott.

The INBC decided to organize a major conference to be held in Mexico City in early February 1984, to which it would invite some of the main institutions who were considering their position on the boycott: grass-roots organizers in the

United States, representatives from the boycott coordination groups in other countries, and health workers and representatives from NGOs in developing countries. The purpose of the meeting was to stimulate support for the boycott by demonstrating that, despite the claims made by Nestlé, people working in the field were still able to find fresh evidence of unacceptable marketing practices. At the same time, the meeting hoped to develop an agreed timetable for an end to the boycott, providing Nestlé met a series of conditions. The meeting also planned to develop a plan of action for pressure on other companies, on governments and health care institutions, to improve the pace of implementation of the code and other necessary changes. IBFAN agreed to co-sponsor the conference, as it was seen as an important opportunity to develop effective strategies for the next phase of the overall campaign.

The INBC drew up a position paper which identified what it saw as the four remaining substantive points of difference between Nestlé's instructions and the international code, and announced at a press conference on 15 December 1983 that if Nestlé 'significantly improves in these areas', the INBC would recommend to the Mexico meeting that the boycott be suspended. The points were that Nestlé should:

- limit free supplies to infants who have a medical need for them (orphans, mothers who are physically incapable of breast feeding);
- stop personal gifts to health workers;
- revise their misleading literature for mothers and health workers to include all information required by the code and to eliminate advertising in written materials given to mothers;
- include health hazard warnings on the labels of infant formulas to specify the hazards of artificial feeding.[48]

Nestlé had consistently refused to negotiate with INBC over the boycott, and there seemed little likelihood that the company would suddenly do so, particularly given the strategy outlined by Pagan to avoid dealing with the leaders of the campaign. However, fate or chance intervened. Douglas Johnson, national chairperson of INFACT, was on his way, by train, from New York to Washington for the press conference to announce the new INBC position. He bumped into Dr Niels Christiansen from Nestlé's Washington Office and outlined the demands, reminded Christiansen that the Mexico meeting was likely to lead to increased boycott activity in other parts of the world, and explained INFACT's plans to open up several more grass-roots campaign centres in the United States in the first few months of 1984. Johnson suggested that Nestlé had better start negotiating fast. Christiansen said he thought it unlikely that Nestlé would be interested, but later that day Johnson received a phone call from Nestlé's Washington office to ask for a meeting to talk over how to proceed.[49]

Nestlé was still reluctant to be seen 'negotiating' with the INBC, so the company approached UNICEF to ask if it would listen to both sides and give guidance as to which it thought was the correct interpretation. UNICEF refused, saying it could not take on such a role if only one party requested it. The INBC was advised of this, and it too requested UNICEF to adjudicate. A series of intensive meetings went on during the early part of January 1984, which were in fact negotiations, although Nestlé would still not describe them in that way.

Nestlé agreed that it would work with a consultant recommended by UNICEF to develop and field-test new educational materials which incorporated all the requirements of the code, and would aim to have such materials ready for distribution by the end of 1984. A similar agreement was reached on labels, although the time-scale was slightly longer, taking between eighteen and twenty-four months to complete. On the question of personal gifts to health workers, it was agreed that these would be limited to inexpensive items of professional utility or educational value and would not carry any product advertising on them. The question of supplies was a major sticking-point and it looked as if the talks would break down over it, until finally, Nestlé agreed that supplies should only be made available to mothers of infants who 'need' to be artificially fed, and also made the amazing concession that it would accept, in advance, a definition of 'need' to be developed by UNICEF.[50] Johnson records:

> When finally, on January 20th, Nestlé made their offer on the supplies policy, we reacted with anxiety, disbelief, relief. It only started to sink in when I got home and started to cry—a mixture of relief, anger, joy, sadness. I also felt sad for the Nestlé officials because they had a sense of defeat by doing the right/ethical thing.[51]

On 25 January 1984, Carl Angst for Nestlé, and William Thompson, stated clerk of the United Presbyterian Church representing the INBC, signed a joint statement (reproduced in Table 7), and the following day a joint press conference was held in Washington to announce to the world that Nestlé had agreed to abide substantially by the international code, that the INBC would recommend to the boycott groups at the Mexico meeting that the boycott be suspended, and that INFACT was immediately suspending the boycott in the United States.[52]

Despite a flurry of long-distance telephone calls, most people involved in the boycott in Europe heard about the details of the agreement from the media. The European boycott groups went to the Mexico meeting with mixed feelings. There was joy about Nestlé finally having negotiated directly with the INBC; hope that the agreement meant a real commitment by Nestlé; but a feeling of anger over the timing of the announcement and the failure of the American groups to consult sufficiently and help the European groups to deal with the deluge of enquiries from the media and others when the news broke. A worry also began to emerge that the agreement would only hold in developing countries: Nestlé's marketing practices in Europe and other industrialized countries did not appear to be covered.

Table 7 Joint statement of INBC and Nestlé

The International Nestlé Boycott Committee has decided to suspend its international boycott of Nestlé Company products.

Over the last decade a controversy has grown over the proper role of infant formula in the Third World, ways of marketing the products there, and the health hazards for infants resulting from improper use.

Much of this controversy has been directed at infant formula manufacturers, and in 1977 INFACT, supported by other organizations, initiated a boycott of the products of Nestlé, the world's largest manufacturer of infant formula.

Out of concern for the health of infants, the World Health Assembly in May,1981 adopted the International Code of Marketing of Breast-milk Substitutes (WHO/UNICEF Code) which provides guidelines to industry, governments, health authorities and non-governmental organizations. Nestlé made a firm commitment to implement that code. Since that time there has been considerable pressure on all infant formula companies to fully comply with the Code, and much progress has been made.

INBC continued to express concern about differences it had with Nestlé's interpretation of some provisions of the Code. In December, 1983 INBC announced four areas of concern: educational material, hazard warnings on labels, gifts to health professionals and free supplies to hospitals. Both Nestlé and INBC sought further guidance from WHO and UNICEF on these areas of the Code.

As a result of intensive discussions involving UNICEF, Nestlé further clarified its policies in these four areas of concern and INBC has recommended a suspension of the boycott.

Both parties praised UNICEF's assistance in clarifying provisions of the Code.

INBC commends Nestlé for taking the leadership role in industry's compliance with the International Code.

Nestlé recognizes and supports the commitment of INBC and its members to safeguard the children of the Third World from hazards related to the inappropriate marketing of infant formula.

Source: *Nestlé News*, January 1984, p. 2.

This was the criticism that had been made by UNICEF nearly a year previously and there were signs that Nestlé had not changed its position. When the company reported on a speech made by UNICEF director James Grant in December 1983, in which he said that he had been told by Nestlé that it 'now supports the WHO Code in all countries', Nestlé added the word 'developing'.[53] Dr Robert Campbell president of the American Baptist Church, and a member of NIFAC, said that Nestlé had told NIFAC that it was 'not willing to unilaterally give up the market in Europe to competition' by applying the code there.[54]

At the Mexico meeting, which involved 120 people from more than fifty organizations in twenty-seven countries, representatives from the INBC found themselves having to 'sell' the agreement in the face of critical scepticism from European and Third World representatives. People from Asia, Latin America, Africa *and* Europe had come to the meeting with fresh examples of promotion by Nestlé and other companies that had been collected only days before. They were sceptical about Nestlé's compliance and were disappointed about the publicity that accompanied the agreement which had suggested that not only was the boycott over, but the entire issue had been resolved. Scepticism about Nestlé's compliance was not helped when forty people at the meeting visited a maternity hospital in Mexico City where the official policy was to encourage breast feeding. Nevertheless, samples of Nestlé's Nan and Wyeth's SMA were found, and although the hospital never requested it, there was sufficient powdered baby milk supplied for every baby. Gay Palmer of BMAC commented:

> There was little doubt as to the sincerity and commitment of the hospital directors to positive breast feeding policies. The discrepancies underline the difficulties of implementation which are common to all hospitals. Many factors are involved, but that of unsolicited supplies being fed into the hospital system is clearly the most insidious and destructive.[55]

After long and involved discussions, the meeting agreed to accept the INBC recommendation that the boycott be suspended for a six-month period to allow for intensive monitoring to establish whether Nestlé was honouring its commitment. If after six months Nestlé was found to be in compliance, the boycott would be officially terminated. If not, the boycott might be resumed, or other pressure brought to bear on the company. The meeting also requested the INBC to resume discussions with Nestlé in an effort to press the company to implement the code in all countries.[56]

The meeting went further and recommended strong actions such as lawsuits or reparation campaigns against companies whose practices endangered infant health; the establishment of an international network of committed health workers to facilitate the exchange of information, strategies and support; and increased support for women-to-women counselling groups and more involvement of women in determining practices related to the reproductive cycle.[57]

Jeanine Velasco, IBFAN's South American coordinator, said the Mexico meeting helped to establish solid links between the concerns of groups in the Third World and those in industrialized countries. 'It has helped to build a solid foundation for future action together to press for changes that will make better infant health a reality'.[58]

In September 1984, the INBC formally met to review Nestlé's compliance with the negotiated agreement. The information from monitoring surveys from around the world showed that Nestlé was still violating the international code. Regular monthly accountability meetings had been held between Nestlé and the INBC since January 1984, and the outcome of these was that Nestlé was visibly *trying* to implement the international code, at least in developing countries, but was faced with production delays on new labels and new materials, and the failure of UNICEF or WHO to come up with any definition over the question of supplies of breast-milk substitutes to hospitals. Thus there were indications of movement by Nestlé, but the actual changes had not yet reached the field.

The willingness of Nestlé to apply the code in Europe was also in some doubt. It was clear that Nestlé had not done everything it had promised, although the fault for that did not necessarily always lie with Nestlé. The INBC representatives had a difficult decision to make. If they recommended a complete end to the boycott, was there any guarantee that Nestlé would continue the process of change? On the other hand, was the failure of Nestlé to meet all the fine print in the agreement a sufficiently strong motivating force to sustain the boycott?

In the end, the balance was swung by action taken by Carl Angst, the executive vice-president of Nestlé. A week before the INBC meeting, Angst sent a letter to Lisa Woodburn, the European coordinator of INBC, in which he stated that although Nestlé could not unilaterally implement the code in Europe because of the strong competition from other companies in that market area, 'the company favours strong national codes' and was willing 'to make this position public'. On the day the INBC meeting began, Angst sent a telex to Douglas Johnson of INFACT stating that the practice of providing mothers with sample packs of infant formula when they leave hospital 'should be stopped . . . I pledge our whole-hearted support to achieve the objective of stopping discharge tins', again with the proviso that it was an industry-wide action.

Such commitments, coming from Angst when he was heavily involved in Nestlé's negotiations to buy the Carnation company, demonstrated to the INBC how important the infant-feeding question was at the very top level of Nestlé's management and how much the company was willing to do to demonstrate its desire to cooperate. Without those commitments, it is quite possible that the boycott would not have been officially called off. With them, the INBC decided to recommend to its member organizations that 'the current international boycott of Nestlé be terminated'.[59]

The news was made public on 4 October 1984 at a joint INBC–Nestlé press conference in Washington. At the press conference, Douglas Johnson said, 'We welcome this decision. While it does not give Nestlé a clean bill of health, the termination of this boycott now gives us the freedom to move on to Nestlé's competitors in the baby milk industry'. At the same time, the INBC pledged to maintain a compliance and accountability process with Nestlé, to hold the company to its commitments, and to seek continuing improvements in marketing practices.[60]

It will be some time before the dust settles and a final analysis can be made about the Nestlé boycott and its conclusion. What is clear is that the ending of the boycott marked an important turning-point in the campaign. It is, perhaps, sad that at the September 1984 INBC meeting the decision to end the boycott was made and then submerged in a welter of hectic activity as the participants divided into small groups to draft press releases, background statements, set up meetings with Nestlé to communicate the decision and make arrangements to inform other campaign groups around the world in advance of the public announcement. It was a full day after the decision was made before the INBC members sat down for a few moments to reflect upon it. The decision represented the death of one strategy and the birth of another. In all societies, such transformations are usually accompanied by rituals or ceremonies to allow for the release of strong emotions. Those rituals are important to prepare people for the future consequences, for the sharing of sorrow and joy, and the re-dedication that provides the basis for continued activity. To have failed to commemorate such a turning-point at the moment of its occurrence was a reflection of the tension that surrounded the decision. The NGOs were entering virgin territory. A negotiating process, traditionally the preserve of the TNCs, had led to a signed agreement. Concessions had been won. It was a demonstration that organized protest, when well focused, could have an impact on powerful institutions and corporations. People could make a difference.

How permanent the change is remains to be seen. Nestlé has reluctantly agreed, after ten years of pressure, to change its marketing practices for a particular product line; it has not substantially altered its corporate policy in other areas. Nestlé, by the way, has a different interpretation of the events that led up to the conclusion of the boycott:

> Nestlé views the situation today as one that has evolved naturally: beginning with our original recommendation to Senator Kennedy that the matter be referred to WHO for determination, to our announcement that we would support the Code from the day it was passed, to our issuance of instructions, our formation of the Muskie Commission, our acceptance of NIFAC recommendations, our consultations with WHO and UNICEF and finally, to our being given specific definitions by UNICEF. To us, this has been a continuity of trying our level best to implement the Code and to actually hammering out exactly what the Code calls for.[61]

This is Nestlé's 'historical perspective' on the issue—a history which omits the role that the *New Internationalist*, War on Want, AgDW, ICCR, INFACT, IBFAN, the INBC and countless other groups, health workers, and even governments played in arriving at the agreement.

Nestlé's continuity of action would probably be understood differently by Louis Macatol if he survives long enough to be able to study it. On 14 November 1983, 15-day old Louis was admitted to Philippine General Hospital with jaundice, chest congestion and a serious cough. Born a healthy, full-term baby, his mother was able to breast feed and planned to do so, until her doctor recommended Nestogen formula. Nestlé had provided the doctor's clinic with large quantities of Nestogen, and the doctor gave a free sample to Louis's mother. She was convinced that Nestogen was better than breast milk and continued to use it even though Louis failed to gain weight and became sick. It was unlikely to be Louis's last trip to hospital, for, by the time he reached five months of age (if he survived that long), the amount of formula he would need would cost the equivalent of his mother's entire income.[62]

Notes

1. INFACT, *Summary Report on the International Nestlé Boycott Meeting, 27-29 May 1983, Gex, France*, Minneapolis, 1983, p. 3.
2. A. Chetley, 'Our campaign has just begun', *Ideas Forum*, 14, UNICEF, 1983.
3. Nestlé, press release, 'Nestlé completes WHO Code implementation process', 16 March 1982.
4. *INFACT Update*, 'Nestlé spreads the word!', May 1982, p. 4.
5. Nestlé, *WHO International Code of Marketing of Breast-milk Substitutes: Instructions to All Companies of the Nestlé Group and to Agents and Distributors Who Market Infant Formula Under Trade Marks Owned by the Nestlé Group*, Washington/Vevey, February 1982.
6. *Sun* (Baltimore), 'Nestlé will obey code on formula', 17 March 1982.
7. *Lancet*, 'Response by Nestlé to the WHO Code on Breast-milk Substitutes', 22 May 1982, p. 197.
8. *INFACT Update*, 'Preliminary analysis: Nestlé's WHO Code interpretation', April 1982, pp. 1–2.
9. Letter from a group of lawyers to Edmund Muskie, Chairman of the Nestlé Infant Formula Audit Commission, 30 June 1982.
10. INFACT, *Baby Bottle Disease: A Call for Action*, Minneapolis, 1982.
11. Letter from James Grant of UNICEF to Rafael Pagan of Nestlé, 10 May 1982; also reported by P. J. Hilts, 'UNICEF hits Nestlé on infant formula', *Washington Post*, 10 June 1982.
12. WHO/UNICEF, *Infant and Young Child Feeding: Notes on the International Code of Marketing of Breast-milk Substitutes*, Geneva, 1982. (They also issued *Comparative Statements of the International Code, the Nestlé instructions and the WHO/UNICEF notes of Clarification*, New York/Geneva, August 1982.)
13. *Aide-mémoire*, prepared by Jonathan Churchill, a lawyer representing the International Nestlé Boycott Committee, of a meeting between himself and J. Paternot, G. Fookes and R. Pagan of Nestlé, held in New York, 20 April 1982, p. 9.
14. Nestlé, *WHO International Code of Marketing of Breast-milk Substitutes: Revised Instructions to Companies of the Nestlé Group and to Agents and Distributors Who Market Infant Formula Under Trade Marks Owned by the Nestlé Group*, Washington/Vevey, October 1982.
15. UNICEF, memo to field directors, 7 February 1983; also reported in *IBFAN News*, 'UNICEF has "substantive differences" with Nestlé', May 1983, p. 2.
16. The other members were Henry Andersen, pastor of Fairmount Presbyterian Church in Cleveland, Ohio; Dr Lewis Barness, professor and chairman of the Department of Paediatrics at the University of South Florida; Robert Campbell, general secretary of the American Baptist Churches of the USA; Dr Omar Fareed, on the American Board of Internal Medicine and medical director of the Carr Foundation; Dr Sheldon Margen, professor of Public Health at the University of California, Berkeley; Vijaya Melnick, special assistant for Policy and Bioethics at the National Institutes on Ageing of the US National Institutes of Health; Mildred Randall, associate professor at the School of Nursing of American University; Philip Wogaman, dean and professor at the Wesley Theological Seminary in Washington, DC. Bishop Ricardo Ramirez, Bishop of Las Cruses, New Mexico also served on the commission for a few months, but subsequently resigned due to an increased workload in his diocese which made it impossible for him to find the necessary time. He was replaced by Fetaui Mata'afa, widow of the first prime minister of Western Samoa, founder and honorary president of the Western Samoa National Council of Women.
17. Tape recording of press conference given by Edmund Muskie, Geneva, 14 May 1982.

18. Tape recording of interview with Edmund Muskie, broadcast on Swiss Radio International, 14 May 1982.
19. NIFAC, *Charter, Articles of Incorporation and By-Laws and Procedures*, Washington, DC, 1982, pp. 15–21.
20. NIFAC, *Fourth Quarterly Report*, Washington, DC, 1983.
21. INFACT, *An INFACT Critique of NIFAC's Fourth Quarterly Report*, Minneapolis, 1983.
22. NIFAC, *Transcript of proceedings of Press Conference with Edmund S. Muskie and Dr Carl Angst*, Washington, DC, 14 October 1982, p. 34.
23. NIFAC, *Transcript of Proceedings of Press Conference with Edmund S. Muskie and Dr Carl Angst*, Washington, DC, 21 April 1983, p. 9.
24. NIFAC, *Transcript* (1983), ibid., p. 26.
25. War on Want, internal memo, May 1982.
26. ICCR, *In Support of the Nestlé Boycott*, New York, February 1983, p. 3.
27. WHO, *Director-General's Report*, Doc. A36/7, Geneva, May 1983.
28. Speech by A. G. Malcolm, Minister of Health for New Zealand, delivered at the 36th WHA, Geneva, May 1983.
29. *INFACT Update*, 'Two victories for Chicago Campaign Centre', May 1983, p. 3.
30. *INFACT Update*, 'Boycott petition drive a great success', June 1983, p. 3.
31. INFACT, *Summary Report*, op. cit. (note 1).
32. *INFACT Update*, 'Auditing the Audit Commission', June 1983, p. 5.
33. E. Marshall, 'Nestlé letter stops NIH talk', *Science*, **219**, 4584, 4 February 1983, p. 469. See also NIFAC, Transcript (1983), op. cit., pp. 38–42.
34. *Guardian*, 'Charity claims it refused $1m offer from Nestlé', 7 February 1983.
35. *Documentation of the Cancellation of Ms Liv Ullmann's Appearance in the TV Show, Ms Ellaneous* (unattributed, but possibly a UNICEF internal memo), Manila, 1983. (According to a report in the *Metro Manila Times*, 6 April 1983, the scriptwriter was the daughter of the head of the advertising agency handling the Filipro account.)
36. *Philippine News*, 'Liv Ullmann was "shocked" for being censored', 6–12 April 1983.
37. *USA Today*, 'Ullmann claims Nestlé kept her off TV show', 23 March 1983.
38. NIFAC, Transcript (1983), op. cit., p. 37.
39. Ibid., p. 38.
40. UNICEF, *Transcript of Press Briefing by Liv Ullmann*, London, 16 June 1983, pp. 8–9.
41. *Bulletin Today*, 'Ullmann charge probed', 1 April 1983; *Metro Manila Times*, 'Multinationals vs breastfeeding (a case of malpractice)', 6 April 1983.
42. J. Melville, 'Which ads are you sold on?', *Guardian*, 21 September 1983, p. 11.
43. S. Nicely, 'Boycott of Nestlé is called victory', *Kansas City Times*, 19 March 1983, p. B1.
44. A. Chetley, op. cit.
45. R. D. Pagan, 'Carrying the fight to the critics of multinational capitalism: think and act politically', speech delivered to the Public Affairs Council, New York, 22 April 1982, reproduced in *Vital Speeches of the Day*, **XLVIII**, 19, 15 July 1982.
46. See F. P. L. Sommerville, 'Support decreases for Nestlé boycott', *Sun* (Baltimore), 25 January 1984, for information on the Church of the Brethren; *Nestlé News*, November and December 1983 issues, for information on the Methodists, Church of the Brethren and Nestlé's response.
47. See *Infact Update*, January 1983, for information about the teachers, and *Nestlé News*, November 1983, for the others.
48. *INFACT News*, winter, 1984, p. 1.
49. Personal communication.
50. See, for example, *Nestlé News*, January 1984; *IBFAN News*, March 1984; *Nestlé Statement*

of Understanding, Washington, DC, Nestlé, 24 January 1984; *INBC Statement of Understanding*, New York, INBC, 25 January 1984; INBC, *Record of Proceedings* (Mexico meeting), Toronto/New York, February 1984.

51. INBC, *Record of Proceedings*, ibid., pp. 6–7.
52. See, for example, N. Dunne and A. McDermott, 'Nestlé boycott could be suspended soon', *Financial Times*, 27 January 1984; *Daily Telegraph*, 'US boycott of Nestlé lifted', 27 January 1984; *Guardian*, 'Nestlé satisfies boycott group', 27 January 1984.
53. *Nestlé News*, December 1983, p. 3.
54. INBC, *Record of Proceedings*, op. cit., p. 17.
55. Ibid., Appendix E.
56. Ibid., p. 30.
57. Statement by conference participants, Mexico City, 6 February 1984.
58. INBC/IBFAN, press statement, Mexico City, 6 February 1984.
59. INBC, *Steering Committee Meeting* (Minutes), New York, 28–29 September 1984, p. 16.
60. INBC, press release, 'Boycott groups launch new strategies for action', 4 October 1984.
61. *Nestlé News*, 'Editorial: historical perspective', February 1984, p. 2.
62. *IBFAN News*, January 1984, p. 4.

10. The lessons of the campaign

> We must ask ourselves radical questions of the kind seldom heard today: what kind of human beings are we, and what do we want to become? The poverty of the marketplace is a wretched visitation, but because it is unlike anything we have ever seen before, we are slow to assess its damage. Epic changes have occurred in our lives, unnoticed. Our will to struggle has been eroded, our experience has been banalised until we doubt the evidence of our own senses and feelings. We must reassert ourselves: define our own needs, not permit them to be implanted in us by agencies that care nothing for human beings. If we cannot respond to the forces that diminish us in the ways we have seen, we shall fall into an underdevelopment yet more abject than any on earth.
>
> Jeremy Seabrook, 'The great consumer swindle', *Guardian*, 21 August 1982

After a decade of campaigning on the infant-feeding issue, the first question which has to be asked is: has anything really changed? The answer is both affirmative and negative.

Yes, there have been changes. A critical awareness has developed amongst many sectors of the world community concerning the need for action to protect infants' rights to adequate nutrition. There is less obvious promotion of bottle feeding in many countries brought about largely through public exposure of the problems, and the consequent public pressure that has been generated in all sectors of the community. An international code exists which provides the minimum standard for marketing behaviour. A handful of governments have adopted it and begun to implement it. A large number are moving in that direction, as well as launching public education campaigns and better training programmes for health workers.

One company, Nestlé, has been forced to negotiate an agreement with pressure groups, whereby it has committed itself to abide by most of the provisions of that code, in most countries. Other companies have made some changes in their marketing strategies, and more appear likely to do so. Two UN agencies, WHO and UNICEF, have increased their stature and legitimacy somewhat in the world community through the efforts they have made on this issue; perhaps UNICEF to a greater extent due to its more sustained efforts to put into practice the internationally-agreed policies that will lead to improvements in maternal and child health.

The issue is slowly becoming an integral part of popular culture—a novel based on it has been written by a Nigerian, a play produced in the United Kingdom, and a calypso record put out in the Caribbean.[1] Health workers have acted in scores of countries to change practices in their own facilities. Dr Natividad Clavano's activities in the Philippines, Dr M. A. Muttalib in Bangladesh, Dr Raj Anand in India, Dr Edgar Rey in Colombia and doctors and health visitors at the Children's Hospital in Sheffield in the United Kingdom, represent only a minute number of possible examples.[2]

A whole host of activities have been undertaken by IBFAN, its affiliate groups and other NGOs to publicize the problem, promote the solutions and press for the necessary changes. The establishment of IBFAN was, itself, a positive step forward, making a new initiative in the way in which NGOs can organize internationally, and serving as a model for similar networks.[3]

Yet, as Willetts points out in an earlier study of the activities of pressure groups on six international issues, an activist would not be satisfied with the progress made, and might even deny that the concept of success is appropriate, given what still remains to be done.[4] The same is true of the infant-feeding issue. There is no denying that there has been tremendous success. Equally, there is no denying that much more needs to be done.

In February 1984, War on Want asked six questions in a campaign update published just after the agreement to suspend the Nestlé boycott had been reached.

— Do health care systems around the world provide adequate encouragement and support for breast feeding in their pre-natal, delivery and post-natal practices?
— Are there adequate social and economic supports for women in place, such as maternity leave, creches and other community and governmental support?
— Are weaning practices appropriate, with adequate emphasis on the use of local food resources?
— Do health workers, parents and the general public receive clear, consistent and correct information, education and training about infant and young child feeding?
— Has the health and social status of women improved so that they are able to play a full and active role in their community and have access to adequate nutrition?
— Is the marketing and distribution of baby milks and weaning foods appropriate, and have all companies and countries implemented the International Code?

The short answer to those questions, which were based on the recommendations of the 1979 WHO/UNICEF meeting on infant and young child feeding, is of course 'no'.

Yet the very success of the campaign, paradoxically, makes it difficult to sustain concerted effort to complete the process of change. In the early 1970s, a ten-minute stroll through a maternity ward, or a few minutes listening to radio would be sufficient to collect enough evidence of heavy-handed commercial promotion of bottle feeding, coupled with inexperienced advice from health workers. Today, that promotion has become more subtle; the message on everyone's lips is: breast feeding is best, even if the practical support to make that happen is still lacking. At the same time, the explanations of the issue have become more complex and the urgency for radical change seemingly less necessary.

Part of the reason for the increased difficulty in keeping the issue on the international agenda is the consistent line put out by the infant-food industry: the controversy is over. One of the first rules of marketing is the need to repeat the essential message—usually the brand name or a snappy catch-phrase—so that it sticks in the mind. From the time ICIFI launched its code in 1975 up to the present, the industry has applied that marketing strategy to its public statements on the issue. At every point of change—the launch of ICIFI, the 1979 WHO/UNICEF meeting, the passage of the international code, the end of the Nestlé boycott—the industry has claimed: we've changed; the controversy has ended.

At first, the statement was treated with the scepticism it deserved, but slowly, like water dripping on stone, the scepticism has been eroded. It is comforting to believe that campaigning efforts have led to success and, therefore, the assurances that the companies now give that they have mended their ways can be used as evidence of success. But it is a double-edged sword. The relaxation of pressure means that further changes are unlikely to occur.

After each of the major turning-points in the campaign, a hiatus set in. Activity slumped, media interest waned, a belief was built up in people's collective consciousness that other topics could now be pursued. The only way to prevent that hiatus was to articulate clearly the next step in the process, having first ensured that the support base for the campaign understood that it was, indeed, *a process*, not a single activity. After the WHO/UNICEF meeting, the next step was obvious: produce a code. Once the code was approved at the WHA, the next step was also fairly obvious, though much more difficult to achieve: implement the code, and bring about the other changes that would reinforce the effectiveness of the code. With the end of the boycott, the opportunity for a single international focus disappeared, and the options for further action have become much more complex.

Other companies can be targeted. Specific countries can be targeted to act as models. Improving the status of women or strengthening work on weaning foods are other strands that could be pursued, as the questions mentioned above indicate. None of these, however, provides the sharp focus that has existed in the past. There is no single enemy, no simple solution—not that there ever was, but by targeting the infant-food industry in general and Nestlé in particular, it was possible to create the illusion of one and, therefore, to concentrate effort and impact.

Thus, in looking back over the decade or so of this campaign, an interesting picture emerges of both the approach needed to present information about a particular issue and the organizational strategies required to ensure that action occurs. With infant feeding, as with most other issues, an extremely complex set of interrelated factors was at play. Initially, these factors were reduced (the industry would say over-simplified) to a single common denominator and a single target: in this case, that the profits of some TNCs were being earned from the ill health or death of infants and the suffering of their parents. Put more bluntly, the problem, cause and solution were reduced to: bottle feeding can kill; TNCs promote bottle feeding; TNCs must stop the promotion.

This is not an argument that is sustainable for very long without buttressing. The industry was quick to retort that its promotion of bottle feeding was not the only cause, or even a significant cause; that the reason bottle feeding kills is due to incorrect preparation or utilization, and that stopping promotion is not the answer.

The industry was not alone in that argument. Dr Charles May, former editor of *Pediatrics*, and a professor in the Department of Pediatrics at the University of Colorado School of Medicine, says:

> The crucial point for pediatricians to realize is that during more than five years of debate and hearings, no substantial, sound, scientific data were ever set forth by the critics of industry or officials of the WHO to support the claim that marketing practices for infant formulas have actually been a significant factor in decline in prevalence of breast feeding in the Third World or anywhere else.[5]

Dr Lewis Barness, former chairman of the American Academy of Pediatrics Committee on Nutrition (CON), and currently a member of the Nestlé Infant Formula Audit Commission, explained that the CON

> feels that the WHO Code attacks a single and perhaps the least important issue, and it is a vendetta against certain infant formulas, namely those which in aspects simulate human milk, and that at best the Code would do little to accomplish the stated aim of fostering breast feeding and at worst, would foster the use of more contaminated, more inadequately prepared, breast milk substitutes, with subsequent increases in infant morbidity and mortality.[6]

Dr A. E. Dugdale, head of the Human Nutrition Research Group of the Department of Child Health at the University of Queensland, said:

> Scientists should view with misgivings some aspects of the WHO Code for the marketing of breast-milk substitutes . . . The Code is based on two assumptions, both testable. The first is that breast feeding is better than other forms of feeding. The other is that mothers, especially those in developing countries are so unable to distinguish the best interests of their infants that they can be deluded by advertising and commercial promotion. These assumptions are obvious simplifications of the real world that mothers inhabit . . . Much of the evidence for and against breast and bottle feeding is of doubtful scientific validity.[7]

An answer to Dugdale's comments (as well, in passing, to the comments of May and Barness) was provided by Drs Derrick and Patrice Jelliffe, with a small selection of scientific facts about the benefits of breast feeding and a convincing argument about the impact of promotion.[8] A further response came from Dr Allan Cunningham, who said Dugdale's call for scientific evidence was 'high-sounding but it comes too late'. He then went on to ask two questions that are pertinent to Dugdale, the other doctors mentioned, and the myriad of other objectors to the campaign for better infant and young child feeding practices. 'Where was he when the marketing juggernaut shifted from the industrial West to the developing nations? Where were physicians 30 years ago when the manufacturers really began their push?'[9] The objectors, of course, do have a point. The control of inappropriate marketing practices is not the complete answer to the problems of infant health. Yet without a starting-point, in this case the activities of the TNCs, it is unlikely that any complex issue would ever be resolved. Sam Shuster notes that 'you never advance a complicated problem with a broad balanced approach as much as with a sharp point, however painfully misplaced'.[10]

The focus on the industry was such a sharp point. Undoubtedly, the argument will continue as to whether or not it was a misplaced attack. In many ways, it is an irrelevant question. If, for a decade or more, the pressure groups had simply stuck to the first premise, and not expanded the debate on the issue, then the comments of May, Barness and Dugdale would warrant attention. However, there has been a gradual unfolding of the whole chain of events, processes, influences and ideologies that affect this subject. This analysis is something which NGOs have been uniquely able to provide. Indeed, the study of the whole chain of production in society is a common factor in many pressure-group activities.[11]

As this unfolding of information continues, the campaign both deepens and broadens. Within IBFAN, there is now a considerable expertise on the intricacies of codes, their benefits and shortcomings; the marketing strategies of TNCs; the influence-peddling activities of TNCs; and the way in which international organizations such as WHO and UNICEF function. It is an almost universal occurrence amongst NGOs that they become specialists in certain fields.[12] Within IBFAN, there are individuals and groups who have more specialist knowledge and understanding of the code and the infant-food industry than officials in many governments. More and more, that specialist knowledge is being called upon by health ministries, as group members are invited to participate on code-drafting committees at the national level. At the same time, the broadening process has led to the identification of other areas that need to be pursued: groups in South East Asia have looked closely at sweetened condensed milk as a risk factor in infant malnutrition; groups in Asia and Europe are studying ways to improve the education and training of health workers; maternity legislation and other social supports for mothers are being examined by groups in all parts of the world; the question of appropriate weaning foods is receiving special attention in Europe and Africa.[13]

> Ted Greiner notes with some satisfaction that this has led to a situation where in recent years, the problem of poor health and high death rates among infants has finally attracted the attention it deserves . . . It sets a precedent in terms of

> coordinated world-wide action in trying to solve a serious problem in which non-governmental organisations played a key role.[14]

The ability of NGOs to play that role depends on their organizational strategies as well as their informational capabilities. The campaign began in a small way, with a few prominent health workers raising questions in the scientific literature and using whatever structures they had available to them to encourage continued discussion. It was at this point a relatively private concern. Paralleling this were the unsung efforts of health workers, individuals and breast-feeding mothers' groups who were trying to motivate mothers to breast feed. This, too, was relatively 'private'—it was not exciting news and therefore received little publicity or prominence. Another characteristic of these activities was the lack of interaction and virtually any coordination. Groups or individuals in one country often were not aware of similar groups in other countries. Dr Thom Kerstiens notes that for NGOs to be effective, they must learn 'to cooperate with each other. This cooperation must be highly flexible and pragmatic, as different situations require.'[15]

The next stage of the campaign, the first public phase, was also largely uncoordinated. IOCU's draft code and its commencement of some research on the issue, the *New Internationalist* and War on Want reports, and the translation by AgDW, all occurred more or less spontaneously. The events were beginning to interconnect: the *New Internationalist* knew about IOCU's efforts; War on Want knew about both; AgDW knew about the three previous initiatives. Each activity reinforced what had gone before, but there was no real unity of purpose. Each activity did have one thing in common: it began from a recognition of an international problem and the links between the causes of the problem and the home country in which the NGO was based. It was the dawning of an international perspective amongst, largely, national organizations.

This also characterized the next stage of development in the campaign as it spread across the Atlantic via ICCR and INFACT. They, too, began activities that followed on from what preceded, recognized the international roots of the problem and looked to see what could be done in a national setting to work for change. At this stage, the campaign was like a relay race, with the baton of action being passed from one group to another.

The beginnings of coalition-building and some true internationalization came with the start of the boycott. INFACT itself was a coalition and organized by gaining the support of other groups. (To a certain extent so, too, was ICCR, although its constituency was limited to church organizations.) When the boycott spread to Canada, and later to other countries, the idea of working together with the same objective, across national boundaries, was introduced into the campaign. The Berne meeting in 1976 was a first stirring of the idea, but little came of it because too few of the groups involved had ever organized internationally.

The international approach was reinforced by the WHO/UNICEF meeting in

1979 and the establishment of IBFAN. The meeting's recommendations and the proposal for a code provided the necessary common focus, but the mechanics were helped by the fact that three of the founding organizations had a strong international perspective. Both Oxfam and War on Want were used to working in partnership with groups in other countries on a project support basis, although neither had really tried to develop international campaign strategies. IOCU was an international NGO. It maintained close contacts with scores of consumer groups around the world and was increasingly developing campaign topics to be taken up jointly.

Thus the cycle was one of isolated, small-scale local activities, followed by more public national activities, then larger-scale international activities—all becoming mutually reinforcing. The local and national activities did not stop; they became stronger. So their strength helped to reinforce the international activities, and vice versa.

With the passage of the code in 1981, it was clear that the emphasis needed to swing back to national activities, but this time with a difference. The campaign had moved up a spiral—it was on a different level or plane. Emphasizing national action did not mean eliminating international action, but simply implied a redefinition of priorities and a shift in the desired flow of information and resources.

By the time the boycott ended in 1984, another shift was beginning: a movement towards more local action and a strengthening of grass-roots organizations. It is conceivable that the pattern established in the infant-feeding campaign is a recurring one and that the ebb and flow amongst the three areas of action will continue. A report in a UNICEF publication describes the role that IBFAN has played and is likely to continue to play in this process.

> In addition to IBFAN's code advocacy, monitoring and breast feeding promotion, the network is a source of information, technical expertise and funding assistance for member organisations. IBFAN's multi-level approach to the infant feeding issue promises to be an effective strategy for eliminating the health problems and infant mortality associated with unnecessary and improper bottle feeding.[16]

As well as an effective strategy for tackling the infant-feeding issue, the organizational pattern of IBFAN has proved to be a model for other groups to follow. A desire to achieve the rational use of medicines, which has been emerging at the local and national levels in several countries around the world, is now being transformed into an international campaign, coordinated by Health Action International (HAI), organized along similar lines to IBFAN. A corresponding concern over the health and environmental hazards of some toxic pesticides led to the formation of the Pesticides Action Network (PAN).

Both these networks have been encouraging strong local, national and regional activity and identifying possible focuses for international pressure. Like IBFAN, they have looked to the UN system for the forum that could provide additional legitimacy for their campaigns and possible international guidelines or standards to be applied in all countries. HAI has been working primarily with WHO and UNCTAD (UN Conference on Trade and Development), while PAN has focused on FAO and the United Nations Environment Programme (UNEP). Both networks have also benefited from resolutions at the UN General Assembly which have led to the establishment of a consolidated list of products that have been banned, restricted or not approved.[17]

In an interesting development during 1984, representatives from several of the networks met together to look at ways of joining forces. The International Coalition for Development Action (ICDA), the European Office of Consumers' Unions (BEUC), PAN, HAI and IBFAN set up a working group particularly to monitor events in Brussels at the EEC and to develop a strong lobby there on the question of hazardous exports.[18]

What has the industry learned?

If NGOs have learned from the experience of the infant-feeding campaign, so, too, have the companies, and not only the infant-food companies. The infant-feeding issue has become essential study matter in management and marketing schools in Europe and the United States. Michael Peretz, executive vice-president of the International Federation of Pharmaceutical Manufacturers' Associations (IFPMA), told Mike Muller that 'Nestlé and the other companies made a lot of mistakes to get to the point they did . . . We won't ever get to that stage if I can help it.'[19]

Within industry federations, individual companies and the whole gamut of public relations firms and business advisory services, a new concept is being developed: issues management. Bob Leaf, of the public relations firm Burson–Marsteller advises that

> companies can't wait for a thing to become an issue and then react. Because then they are on the defensive. The key to the 80s will be defining the issues before they have an impact on you so that you can diffuse them, be prepared to have an action plan when something comes up rather than having to act hurriedly under attack.[20]

As long ago as 1978, *Business International* was warning of the dangers of failing to act before an issue became entrenched. In a survey of 100 major American and European firms, they found most did not become involved in the public policy process until it was well advanced, usually at a stage when it was too late to have any impact on impending regulations. Some of the reasons for this included:

— *Business's perception of its role*. Companies have historically seen as their only function the creation of wealth.

— *Insularity*. Companies have been largely protected from the public policy process because governments tend to share business's own view of its mission.
— *Wishful thinking*. Many executives have harboured the hope that, whatever the issue, it might just go away.
— *Lack of commitment*. Boards and chief executive officers have been slow to accept the qualitative changes in the 'contract' between business and society. Acceptance, where it exists, is at an intellectual, pragmatic level, not at the emotional level required for individuals in the company to respond to change.
— *Organizational structure*. The absence of emotional acceptance of the process of change generally prevents corporations from instituting the dramatic organizational shifts necessary to meet the expectations of a changing world.[21]

Four years later, the magazine provided a checklist of six lessons that could be learned from the mismanagement of the infant-feeding controversy.[22] Top of the list was the instruction to ensure that all staff take seriously any voluntary codes of practice, ethics or marketing. This was an appeal to get across the emotional, and not simply the intellectual, importance of such measures. It was fair warning that simply paying lip-service to voluntary controls was a recipe for controversy.

Business International also stressed the need to learn how products are used, preferably before the critics do. This was a clear statement that it was no longer an acceptable argument to say that if all the instructions are followed carefully, the product is safe—particularly if, as in the case of infant-feeding products, the likelihood of that happening is remote, given the economic and social conditions of the end user. Companies were urged to avoid misleading appeals to the modern way of life in promotion campaigns, and to note that an increasingly important theme in much criticism was that TNCs were exporting inappropriate consumption patterns to developing countries. Full disclosure of information about health and safety in the use of the product was advised on all labels and instruction materials. At the same time, companies were advised to involve identifiable constituents in the exporting countries, such as shareholders and workers, in frank discussions on the relative risks and benefits involved in products. This was an appeal to get in early as a means of protecting the image of the company. Finally, companies were advised to address the issue, not the critics. It was noted that, although it might be tempting to expose critics in terms of their background or financial support, such attacks would generally backfire.

In summary, the proposals set out by *Business International* were virtually the opposite of Nestlé's (and the other infant-food companies') early approach to the infant-feeding issue. One of the chief architects of the shift in Nestlé's policy, Rafael Pagan, president of the Nestlé Coordination Centre for Nutrition, shares the belief that Nestlé got it wrong at first. He claims that the fact a boycott occurred 'represents a major failure by a large multinational company to identify legitimate concerns early enough, compounded by inadequate communication to influential opinion leaders of the policy responses actually made by the company'.[23]

Pagan suggests that it is important for companies to go beyond issues management. He would like to see TNCs becoming actively involved in efforts to change attitudes both within the business community and in the wider world. Some of his suggestions for change within industry offer hope: he wants to see business become less secretive and to give clear reasons for its actions; to acknowledge that where business is granted the 'privilege to create wealth and profits' it must also accept a broad array of responsibilities to the community and the world.

Pagan's recipe for changing the perceptions of the larger world involve corporations in thinking and acting 'politically'. That begs the question about whether they have not acted politically in the past. Clearly, there are vivid examples of the political action of TNCs, some of which have been cited in earlier chapters. However, Pagan is calling for more conscious political action. The ideology of such action is obvious:

> We must affirm the community of interest of all wealth-creating institutions, large and small, private and government, national and multinational—in short, the pluralism and variety of the human condition as exemplified by democracy and the free marketplace of commerce and ideas. Multinational capitalism *must never be seen* as an overbearing rival of small-scale local interests, or of national and tribal feeling [emphasis added].[24]

The carefully-chosen wording leaves some doubt about the sincerity of the claims that real changes will be made. One of Pagan's major strategies for political action rests on the idea of coalition-building amongst the business community. He says, 'if activists with radically different world views and goals can cooperate, so must we'.[25] He believes that industries should develop their own 'NGOs' and ensure that they gain representation 'at every possible UN agency' as a means of dealing 'constructively and effectively' with the international mood for regulation. He claims the activists gained 'a political head start' in this, but if companies learn to think and act politically, they can ultimately defeat activist critics.

Certainly, the infant-food industry is pursuing Pagan's advice. After ICIFI was refused representational status with WHO for the third time in 1983, it sank rapidly into oblivion. From its ashes, there rose a new, and potentially stronger body, the IFM (International Association of Infant Food Manufacturers). Formed in mid-1983, it has thirty-one member companies and is a subsidiary body of the International Society of Dietetic Food Industries (ISDI). Beginning in October 1983, the IFM entered into discussions with WHO about gaining official NGO status, hoping that it could use the long, informal working relationship period established by ICIFI as a means of avoiding the normal two-year minimum waiting period before the application could be approved. It was expected that in August 1985 IMF would officially apply for NGO status with WHO; however, this has not happened. One of the worries that IMF had about the process was the attitude of IBFAN, which had been so successful in lobbying against ICIFI's three applications. IFM told WHO it was not prepared even to consider applying unless it considered that its chances of success were reasonably good.[26]

It is, perhaps, a glowing compliment to the effectiveness of IBFAN that a major multinational corporation such as Nestlé has decided to copy some of IBFAN's informational and organizational strategies. As far back as 1980, Nestlé and ICIFI recognized the need to involve 'grass-roots' organizations.[27] After IBFAN began producing its regular reports on company violations of the 1979 WHO/UNICEF recommendations, which included photographs of actual advertisements or posters plus powerful graphics to convey the simple message that the companies were breaking the rules, Nestlé responded with a well-designed rebuttal which included photographs of non-offensive posters, new labels, repainted delivery vans and some effective typography to maintain the reader's interest.[28] In late 1983, the company began producing a monthly newsletter, *Nestlé News*, modelled on *IBFAN News*, but with better design and graphics. The stories were short and punchy, and for the most part, played up news about organizations that were withdrawing from the boycott. Ostensibly, it was a publication to provide information 'about human nutrition'. In reality, it was a public relations document displaying the positive aspects of the corporation. It was effective.

Nestlé News also provided the company with a vehicle with which it could disseminate a new view of the history of the campaign. In an editorial in the February 1984 issue, Nestlé gave its 'historical perspective' on the issue—one that omitted to mention the role played by the various pressure groups over the decade, and that made it appear as if Nestlé was the driving-force behind the changes that had occurred internationally.

The future

A war of words and a war of interpretations will undoubtedly continue for some time between industry officials and the pressure groups as the practical work of translating international guidelines into day-to-day marketing instructions lumbers on. But more importantly, in hospitals, clinics and communities, the efforts of the unsung heroes and heroines—the front-line health workers, the mothers' support groups, the community workers—will continue. It is here that the real battle is being fought. In another ten years, it may be possible to declare a 'winner', or rather a community of winners—adequately nourished mothers and infants.

What will also continue is the effort to provide a voice for the so-called 'third system'—the people's movements, pressure groups or NGOs that represent specific onstituencies or provide the momentum for action on specific issues. Governments of all political colours are often characterized as moribund bureaucracies, slow to respond to changes in society. Similarly, the other major actors on the world stage, the TNCs, are characterized as only responding to those changes if there is a profit to be made in doing so. They, too, are beset with their own bureaucracies, usually somewhat more flexible than those of government. Between these two giants lie the intergovernmental agencies, characterized as part-mediator, part-sounding-board, part-think-tank, part-advisory service. The intergovernmental agencies, too, have their own ponderous bureaucracies. Left to these three groupings, the wheels of change would grind excrutiatingly slowly.

The third system is the driving-force for change. Loosely structured, flexible to the extreme, it is a wild, exciting concoction of ideas, strategies and plans that range anywhere between sheer madness to absolute genius. The pressure groups do not always have the solutions. (Neither do governments, TNCs or the international agencies.) But they have the energy and determination to confront situations and, often, force a search for solutions.

The pressure groups are able to inject a large dose of reality into what are otherwise sterile discussions about the way to organize an ideal world. That reality is often shocking, often unpalatable to the ruling elites at the top of the world's various decision-making structures, whether they be national governments, TNCs or international agencies. The claims and demands of the pressure groups come forward with powerful emotional force, often with anger, frequently with frustration. There are reasons for that anger. The day-to-day sight of the victims of poverty, repression and violence has a politicizing effect. The pressure groups are talking about real lives, real people. They can put faces and names to the faceless statistics.

Throughout the infant-feeding campaign the emotional response has led to moments of great courage amongst individuals and groups. It took courage for Dr Clavano to tear down the bottle-feeding posters in her hospital in the Philippines and replace them with breast-feeding posters in the face of opposition from some of her colleagues, the infant-food industry and a *laissez-faire* government. It took courage for a group of breast-feeding mothers to take their babies into a leading London department store in protest over a woman having been removed from its restaurant for quietly breast feeding her baby while she herself ate lunch. It took courage for the Indian Academy of Paediatrics to turn down offers of funding from the pharmaceutical and infant-food companies to avoid any possible conflict of interest. These were acts that went against the dominant flow of society. They were acts of defiance, moments of catalytic change.

During the campaign, it has also been possible to identify organizations and individuals who have acted with 'discourage'. These include governments who are reluctant to act to change the status quo; scientists who claim that all the possible avenues of research have not been explored, therefore it is too early to make a decision about what to do; people who hide behind the thought that the issue is so complex, and only part of a much larger world malaise, that it is pointless even to try to resolve it.

The infant-feeding issue is a political issue, not party political, but *realpolitik*. In order for long-term change to be sustained, then existing structures must be questioned and, in some cases, questioned bluntly. Inevitably, that creates friction.

A dichotomy appears between those who are in favour of change and those who are not; except the dichotomy is usually blurred by the use of carefully-worded phrases. Few people willingly admit being against change in an emotive situation. Instead they will agree that change is needed and argue about how it should be achieved, a delaying tactic that ensures that change is slow and less painful to them.

Throughout the campaign, those distinctions have been exposed. Confrontation has been used frequently, and has enraged not only the industry but more moderate supporters of the basic cause. At the same time, tactful diplomacy has been under way, the give-and-take of compromise and consensus. Without the confrontation, the compromises would be less palatable to the people most at risk, the consensus impossible to achieve.

Dr F. G. Joseph, formerly with the Public Health Department in Kenya, argues that 'antagonism and rancour on either side (the company's or the health protestors') will not contribute to the reduction of infant mortality'.[29] On one level he is right: antagonism *alone* will not improve infant health. However, it is a necessary ingredient in the final prescription. Without the input of the pressure groups, often antagonistic, this issue would still not be on the international agenda. Governments or individual health workers would not have monitored the industry's voluntary code of ethics, so there would be little likelihood that an international code would have been developed. A decade of activity would have been lost and, during that time, so would many lives. Bottle feeding would be more entrenched and harder to fight against. The pressure groups' activities have not solved the problem, but they have established a climate of urgency, which has contributed significantly to the process for its resolution.

IBFAN and the individual groups in the network have been flexible in their approach. They have worked within the existing national and international structures. But where those structures have proved to be clumsy, intransigent or in opposition to change, they have moved beyond them to organize informally to achieve results. To some, such an approach smacks of anarchy, terrorism and the downfall of democracy. It suggests that articulate and manipulative minorities will be able to influence unduly decisions affecting the majority of the populace. In fact, it could equally be argued that the current system operates in precisely that way: an elite and privileged minority with access to education, finance and authority controls the major decision-making processes and moulds them so that they reinforce the existing system.

Toffler claims that the concept of majority rule is becoming increasingly obsolete. He calls for new approaches that provide for a democracy of minorities and the development of imaginative new institutions that will allow for the transfer of decision-making to its appropriate level. Many health and environmental issues are global issues and should be decided as such. Some decisions need to be made at a national level; others at a regional or local level within a country. He calls for the unleashing of a wide-ranging public debate to generate imaginative proposals for political restructuring.[30]

In some small ways, the development of the infant-feeding campaign, the organization of IBFAN and the role it has played in developing cooperative links at all levels and in all societies, are precursors to what Toffler is suggesting. IBFAN is an experiment in a new way of coping with huge, complex world problems. Like many experiments, it has design flaws and imperfections. But according to Anwar Fazal, president of the International Organization of Consumers Unions (IOCU), the campaign and the founding of IBFAN 'was a watershed for international consumerism'. He noted that it provided a learning experience for those involved about the UN and the interaction between TNCs, the UN and individual governments; offered an opportunity for diverse groups to work together, pooling expertise and resources; and showed how the third system, the citizens of the world, could unite to change irresponsible and inappropriate technology.[31]

Notes

1. D. Fulani, *The Fight For Life*, London, Hodder & Stoughton, 1982. The novel is one of a series of books on contemporary issues designed for new readers in West Africa; S. Wilmer, *The Baby Killer*, a play first produced at the Warehouse Theatre in Croydon, London in 1982, based on a fictional, though true-to-life, health-care facility in Zimbabwe shortly after independence, when a milk company tries to move into a new market; the record was made by a leading Caribbean calypso artist for the Trinidad carnival in 1983 and proved very popular.
2. See, for example, N. Clavano, 'The results of a change in hospital practices: a paediatrician's campaign for breast-feeding in the Philippines', *Assignment Children*, UNICEF, 55/56, 1981. After forbidding the promotion of bottle feeding in the hospital, breast feeding increased by 135 per cent, diarrhoea decreased by 94 per cent, infant mortality decreased by 95 per cent. Dr Muttalib's story is told in A. Chetley, 'There is *no* substitute for mother's milk', *World Health*, February/March 1983. Dr Anand's success is reported in the *Assignment Children*, op. cit. Dr Rey's story is summarized in O. Gillie, 'Nestling into life', *Sunday Times*, 16 December 1984. By encouraging breast feeding for premature babies, he was able to bring their death rate down from 70 per cent to just 5 per cent. The Sheffield story is told in O. Gillie (untitled), *Good Housekeeping*, February 1984, p. 91. The encouragement of breast feeding was calculated to prevent the death of twenty babies through Sudden Infant Death Syndrome.
3. Subsequent networks that have been developed include Health Action International (HAI), formed in 1981 to work on the pharmaceutical issue; the Pesticides Action Network International (PAN), also formed in 1981 to campaign on pesticides; Consumer Interpol, set up in 1982, to act as an international warning system about hazardous products moving in world trade.
4. P. Willetts (ed.), *Pressure Groups in the Global System*, London, Frances Pinter, 1982, p. 195.
5. C. D. May, 'The "infant formula controversy": a notorious threat to reason in matters of health', *Pediatrics*, **68**, 3, September 1981, p. 429.
6. L. A. Barness, 'Committee on Nutrition and the WHO Code of Marketing Breastmilk Substitutes', *Pediatrics*, **68**, 3, September 1981, p. 431.
7. A. E. Dugdale, 'Breast-milk policy should be more scientific', *World Health Forum*, **4**, 1983, p. 214.

8. D. B. Jelliffe and E. F. P. Jelliffe, 'Breast-milk policy', *World Health Forum*, **5**, 1984, pp. 37–8.
9. A. S. Cunningham, 'Breast-milk policy', *World Health Forum*, **5**, 1984, p. 39.
10. S. Shuster, 'Under pressure', *World Medicine*, 6 March 1982, p. 33.
11. An example of this approach can be found in the operations of the French group, Solagral, active on food issues. Solagral follows a fourfold approach to the subjects it tackles: participatory research; analysis of production chains; networking; research into alternatives or solutions. In this way, the analysis and research becomes a means of increasing the understanding of those involved in a production process and empowers them with the information necessary to begin to consider feasible alternative methods. This is further explained in *Report on the Food Strategy Workshop*, published by UN-NGLS and Solagral, June 1983, available from NGLS, Geneva. Another instance of this occurred in the United Kingdom at a War-on-Want-sponsored day school on pharmaceuticals, when all fifty participants divided into small groups and worked through an exercise of looking at the supply chain of drugs from the manufacturer to the consumer, identifying the various steps along the chains, the influences that act upon the chain, and the possible points of intervention to improve the final result. One interesting point was that all the groups concluded that concerns over health did not necessarily figure in the chain except as a side issue. (See *Campaigning on Pharmaceuticals*, Manchester, War on Want, 18 February 1984.)
12. See Willetts, op. cit., pp. 183 and 185–7, for an extension of this argument.
13. Weaning foods have begun to receive universal attention, following a resolution (WHA 37/30) passed at the 1984 World Health Assembly, which noted that '. . . many products unsuitable for infant feeding are being promoted for this purpose in many parts of the world, and that some infant foods are being promoted for use at too early an age, which can be detrimental to infant and young child health'. The resolution called on member states to act on this problem, for WHO to support them on this, and to report back in 1986 on progress, together with recommendations for any other measures that might be needed. The resolution prompted panic within the WHO secretariat, nervous about how the United States would react. The original sponsor of the resolution, a delegate from Qatar, was subjected to constant pressure from the secretariat to tone down the text. When it was subsequently strengthened as a result of an amendment from Uganda, a carefully stage-managed debate forced the matter into a drafting group. At that group, an assistant director-general of WHO argued strongly for a go-slow policy and a moderate resolution that made no call for action. This was resisted by the delegates. In the end, the resolution was adopted without a vote, although it prompted an emotional outburst from an American delegate, while still in committee stage. Neil Boyer, of the US State Department, claimed the resolution 'would push the World Health Organization into more controversy and diminish international support for WHO'. He protested that this was another example of WHO interfering with private enterprise and that 'WHO should not get involved in regulating marketing practices around the world'. He said the new resolution would lead to a renegotiation of the code and a widening of its scope. IBFAN was instrumental in providing information about some promotional practices for weaning foods in Africa that had occurred only weeks before the assembly, including identifying many products on the market totally unsuitable for infants. This information helped strengthen the resolve of delegates who called for action. For further details, see 'US Government objects to WHA resolution', *IBFAN News*, May/June 1984; A. Chetley, 'Infant feeding: US again says no', *Health Now*, 5/6, 17 May 1984.
14. T. Greiner, 'The WHO International Code of Marketing of Breast-milk Substitutes: a time for reevaluation', *IFDA Dossier*, 31, October/November 1982, p. 77.

15. T. Kerstiens, 'Are NGOs in?', *Ideas Forum*, UNICEF, 8, 1982/1, pp. 2–3.
16. UNICEF, 'The International Baby Food Action Network', *Ideas Forum*, 13, 1983/2.
17. On 17 December 1982, the General Assembly approved, with 146 votes in favour, no abstentions, and once again the United States casting the sole negative vote, Resolution No. 37/137 calling for such a list. The first edition was issued at the end of 1983, with a revised version appearing in mid-1984—United Nations, *Consolidated List of Products Whose Consumption and/or Sale Have Been Banned, Withdrawn, Severely Restricted or Not Approved by Governments*, New York, DIESA/WP/1, July 1984.
18. J. Lemmers, 'Issue networks join forces', *ICDA News*, March 1984, p. 1.
19. M. Muller, *The Health of Nations*, London, Faber & Faber, 1982, p. 212.
20. C. Pfaff, 'New threats to multinationals', *Advertising Age/Europe*, December 1981, p. 16.
21. *Business International*, 'Why MNCs fail to act before emerging issues arouse regulatory fervor', 17 November 1978, p. 367.
22. *Business International*, 'Lessons from the anti-infant formula campaign', 5 February 1982.
23. R. D. Pagan, 'Resolving the Nestlé boycott: 1977–1984', *IPRA Review*, August 1984, pp. 21–6.
24. R. D. Pagan, 'Carrying the fight to the critics of multinational capitalism', speech delivered to the Public Affairs Council, New York, 22 April 1982, published in *Vital Speeches of the Day*, **XLVIII**, 19, 15 July 1982.
25. Ibid.
26. Various WHO and IFM internal memos (1983–4).
27. See BMAC, *Nestlégate: Secret memo reveals corporate cover-up*, Cambridge, Mass., 1981. In the memo, Ernest Saunders of Nestlé talks about the success the critics are having in mobilizing local activities and suggests that corporate staff should try to develop a grass-roots approach as well.
28. Nestlé, *Marketing Infant Formula in the Third World: How Nestlé Has Followed the WHO/UNICEF October 1979 Recommendations*, Vevey, April 1981.
29. F. G. Joseph, 'Putting the bottle in its proper place', *Development Forum*, July/August 1982, p. 7.
30. A. Toffler, *The Third Wave*, London, Pan, 1981, Ch. 28.
31. A. Fazal, 'The third force', *Multinational Monitor*, **1**, 9, October 1980, pp. 15–16.

Appendix 1: The International Code of Marketing of Breast-milk Substitutes

The Member States of the World Health Organization:

Affirming the right of every child and every pregnant and lactating woman to be adequately nourished as a means of attaining and maintaining health;

Recognizing that infant malnutrition is part of the wider problems of lack of education, poverty, and social injustice;

Recognizing that the health of infants and young children cannot be isolated from the health and nutrition of women, their socio-economic status and their roles as mothers;

Conscious that breast-feeding is an unequalled way of providing ideal food for the healthy growth and development of infants; that it forms a unique biological and emotional basis for the health of both mother and child; that the anti-infective properties of breast milk help to protect infants against disease; and that there is an important relationship between breast-feeding and child-spacing;

Recognizing that the encouragement and protection of breast-feeding is an important part of the health, nutrition and other social measures required to promote healthy growth and development of infants and young children; and that breast-feeding is an important aspect of primary health care;

Considering that when mothers do not breast-feed, or only do so partially, there is a legitimate market for infant formula and for suitable ingredients from which to prepare it; that all these products should accordingly be made accessible to those who need them through commercial or non-commercial distribution systems; and that they should not be marketed or distributed in ways that may interfere with the protection and promotion of breast-feeding;

Recognizing further that inappropriate feeding practices lead to infant malnutrition, morbidity and mortality in all countries, and that improper practices in the marketing of breast-milk substitutes and related products can contribute to these major public health problems;

Convinced that it is important for infants to receive appropriate complementary foods, usually when the infant reaches four to six months of age, and that every effort should be made to use locally available foods; and convinced, nevertheless, that such complementary foods should not be used as breast-milk substitutes;

Appreciating that there are a number of social and economic factors affecting breast-feeding, and that, accordingly, governments should develop social support systems to protect, facilitate and encourage it, and that they should create an environment that fosters breast-feeding, provides appropriate family and community support, and protects mothers from factors that inhibit breast-feeding;

Affirming that health care systems, and the health professionals and other health workers serving in them, have an essential role to play in guiding infant feeding practices, encouraging and facilitating breast-feeding, and providing objective and consistent advice to mothers and families about the superior value of breast-feeding, or, where needed, on the proper use of infant formula, whether manufactured industrially or home-prepared;

Affirming further that educational systems and other social services should be involved in the protection and promotion of breast-feeding, and in the appropriate use of complementary foods;

Aware that families, communities, women's organizations and other nongovernmental organizations have a special role to play in the protection and promotion of breast-feeding and in ensuring the support needed by pregnant women and mothers of infants and young children, whether breast-feeding or not;

Affirming the need for governments, organizations of the United Nations system, nongovernmental organizations, experts in various related disciplines, consumer groups and industry to cooperate in activities aimed at the improvement of maternal, infant and young child health and nutrition;

Recognizing that governments should undertake a variety of health, nutrition and other social measures to promote healthy growth and development of infants and young children, and that this Code concerns only one aspect of these measures;

Considering that manufacturers and distributors of breast-milk substitutes have an important and constructive role to play in relation to infant feeding, and in the promotion of the aim of this Code and its proper implementation;

Affirming that governments are called upon to take action appropriate to their social and legislative framework and their overall development objectives to give effect to the principles and aims of this Code, including the enactment of legislation, regulations or other suitable measures;

Believing that, in the light of the foregoing considerations, and in view of the vulnerability of infants in the early months of life and the risks involved in inappropriate feeding practices, including the unnecessary and improper use of breast-milk substitutes, the marketing of breast-milk substitutes requires special treatment, which makes usual marketing practices unsuitable for these products;

Therefore:

The Member States hereby agree the following articles which are recommended as a basis for action.

Article 1. Aim of the Code

The aim of this Code is to contribute to the provision of safe and adequate nutrition for infants, by the protection and promotion of breast-feeding, and by ensuring the proper use of breast-milk substitutes, when these are necessary, on the basis of adequate information and through appropriate marketing and distribution.

Article 2. Scope of the Code

The Code applies to the marketing, and practices related thereto, of the following products: breast-milk substitutes, including infant formula; other milk products, foods and beverages, including bottle-fed complementary foods, when marketed or otherwise represented to be suitable, with or without modification, for use as a partial or total replacement of breast-milk; feeding bottles and teats. It also applies to their quality and availability, and to information concerning their use.

Article 3. Definitions

For the purposes of this Code:

'*Breast-milk substitute*' means any food being marketed or otherwise represented as a partial or total replacement for breast-milk, whether or not suitable for that purpose.

'*Complementary food*' means any food, whether manufactured or locally prepared, suitable as a complement to breast-milk or to infant formula, when either becomes insufficient to satisfy the nutritional requirements of the infant. Such food is also commonly called 'weaning food' or 'breast-milk supplement'.

'*Container*' means any form of packaging of products for sale as a normal retail unit, including wrappers.

'*Distributor*' means a person, corporation or any other entity in the public of private sector engaged in the business (whether directly or indirectly) of marketing at the wholesale or retail level a product within the scope of this Code. A 'primary distributor' is a manufacturer's sales agent, representative, national distributor or broker.

'*Health care system*' means governmental, nongovernmental or private institutions or organizations engaged, directly or indirectly, in health care for mothers, infants and pregnant women; and nurseries or child-care institutions. It also includes health workers in private practice. For the purposes of this Code, the health care system does not include pharmacies or other established sales outlets.

'*Health worker*' means a person working in a component of such a health care system, whether professional or non-professional, including voluntary, unpaid workers.

'*Infant formula*' means a breast-milk substitute formulated industrially in accordance with applicable Codex Alimentarius standards, to satisfy the normal nutritional requirements of infants up to between four and six months of age, and adapted to their physiological characteristics. Infant formula may also be prepared at home, in which case it is described as 'home-prepared'.

'*Label*' means any tag, brand, mark, pictorial or other descriptive matter, written, printed, stencilled, marked, embossed or impressed on, or attached to, a container (see above) of any products within the scope of this Code.

'*Manufacturer*' means a corporation or other entity in the public or private sector engaged in the business or function (whether directly or through an agent or through an entity controlled by or under contract with it) of manufacturing a product within the scope of this Code.

'*Marketing*' means product promotion, distribution, selling, advertising, product public relations, and information services.

'*Marketing personnel*' means any person whose functions involve the marketing of a product or products coming within the scope of this Code.

'*Samples*' means single or small quantities of a product provided without cost.

'*Supplies*' means quantities of a product provided for use over an extended period, free or at a low price, for social purposes, including those provided to families in need.

Article 4. Information and education

4.1 Governments should have the responsibility to ensure that objective and consistent information is provided on infant and young child feeding for use by families and those involved in the field of infant and young child nutrition. This responsibility should cover either the planning, provision, design and dissemination of information, or their control.

4.2 Informational and educational materials, whether written, audio, or visual, dealing with the feeding of infants and intended to reach pregnant women and mothers of infants and young children, should include clear information on all the following points: (a) the benefits and superiority of breast-feeding; (b) maternal nutrition, and the preparation for and maintenance of breast-feeding; (c) the negative effect on breast-feeding of introducing partial bottle-feeding; (d) the difficulty of reversing the decision not to breast-feed; and (e) where needed, the proper use of infant formula, whether manufactured industrially or home-prepared. When such materials contain information about the use of infant formula, they should include the social and financial implications of its use; the health hazards of inappropriate foods or feeding methods; and, in particular, the health hazards of unnecessary or improper use of infant formula and other breast-milk substitutes. Such materials should not use any pictures or text which may idealize the use of breast-milk substitutes.

4.3 Donations of informational or educational equipment or materials by manufacturers or distributors should be made only at the request and with the written approval of the appropriate government authority or within guidelines given by governments for this purpose. Such equipment or materials may bear the donating company's name or logo, but should not refer to a proprietary product that is within the scope of this Code, and should be distributed only through the health care system.

Article 5. The general public and mothers

5.1 There should be no advertising or other form of promotion to the general public of products within the scope of this Code.

5.2 Manufacturers and distributors should not provide, directly or indirectly, to pregnant women, mothers or members of their families, samples of products within the scope of this Code.

5.3 In conformity with paragraphs 1 and 2 of this Article, there should be no point-of-sale advertising, giving of samples, or any other promotion device to induce sales directly to the consumer at the retail level, such as special displays, discount coupons, premiums, special sales, loss-leaders and tie-in sales, for products within the scope of this Code. This provision should not restrict the establishment of pricing policies and practices intended to provide products at lower prices on a long-term basis.

5.4 Manufacturers and distributors should not distribute to pregnant women or mothers of infants and young children any gifts of articles or utensils which may promote the use of breast-milk substitutes or bottle-feeding.

5.5 Marketing personnel, in their business capacity, should not seek direct or indirect contact of any kind with pregnant women or with mothers of infants and young children.

Article 6. Health care systems

6.1 The health authorities in Member States should take appropriate measures to encourage and protect breast-feeding and promote the principles of this Code, and should give appropriate information and advice to health workers in regard to their responsibilities, including the information specified in Article 4.2.

6.2 No facility of a health care system should be used for the purpose of promoting infant formula or other products within the scope of this Code. This Code does not, however, preclude the dissemination of information to health professionals as provided in Article 7.2.

6.3 Facilities of health care systems should not be used for the display of products within the scope of this Code, for placards or posters concerning such products, or for the distribution of material provided by a manufacturer or distributor other than that specified in Article 4.3.

6.4 The use by the health care system of 'professional service representatives', 'mothercraft nurses', or similar personnel, provided or paid for by manufacturers or distributors, should not be permitted.

6.5 Feeding with infant formula, whether manufactured or home-prepared, should be demonstrated only by health workers, or other community workers if necessary; and only to the mothers or family members who need to use it; and the information given should include a clear explanation of the hazards of improper use.

6.6 Donations or low-price sales to institutions or organizations of supplies of infant formula or other products within the scope of this Code, whether for use in the institutions or for distribution outside them, may be made. Such supplies should only be used or distributed for infants who have to be fed on breast-milk substitutes. If these supplies are distributed for use outside the institutions, this should be done only by the institutions or organizations concerned. Such donations or low-price sales should not be used by manufacturers or distributors as a sales inducement.

6.7 Where donated supplies of infant formula or other products within the scope of this Code are distributed outside an institution, the institution or organization should take steps to ensure that supplies can be continued as long as the infants concerned need them. Donors, as well as institutions or organizations concerned, should bear in mind this responsibility.

6.8 Equipment and materials, in addition to those referred to in Article 4.3, donated to a health care system may bear a company's name or logo, but should not refer to any proprietary product within the scope of this Code.

Article 7. Health workers

7.1 Health workers should encourage and protect breast-feeding; and those who are concerned in particular with maternal and infant nutrition should make themselves familiar with their responsibilities under this Code, including the information specified in Article 4.2.

7.2 Information provided by manufacturers and distributors to health professionals regarding products within the scope of this Code should be restricted to scientific and factual matters, and such information should not imply or create a belief that bottle-feeding is equivalent or superior to breast-feeding. It should also include the information specified in Article 4.2.

7.3 No financial or material inducements to promote products within the scope of this Code should be offered by manufacturers or distributors to health workers or members of their families, nor should these be accepted by health workers or members of their families.

7.4 Samples of infant formula or other products within the scope of this Code, or of equipment or utensils for their preparation or use, should not be provided to health workers except when necessary for the purpose of professional evaluation or research at the institutional level. Health workers should not give samples of infant formula to pregnant women, mothers of infants and young children, or members of their families.

7.5 Manufacturers and distributors of products within the scope of this Code should disclose to the institution to which a recipient health worker is affiliated any contribution made to him or on his behalf for fellowships, study tours, research grants, attendance at professional conferences, or the like. Similar disclosures should be made by the recipient.

Article 8. Persons employed by manufacturers and distributors

8.1 In systems of sales incentives for marketing personnel, the volume of sales of products within the scope of this Code should not be included in the calculation of bonuses, nor should quotas be set specifically for sales of these products. This should not be understood to prevent the payment of bonuses based on the overall sales by a company of other products marketed by it.

8.2 Personnel employed in marketing products within the scope of this Code should not, as part of their job responsibilities, perform educational functions in relation to pregnant women or mothers of infants and young children. This should not be understood as preventing such personnel from being used for other functions by the health care system at the request and with the written approval of the appropriate authority of the government concerned.

Article 9. Labelling

9.1 Labels should be designed to provide the necessary information about the appropriate use of the product, and so as not to discourage breast-feeding.

9.2 Manufacturers and distributors of infant formula should ensure that each container has a clear, conspicuous, and easily readable and understandable message printed on it, or on a label which cannot readily become separated from it, in an appropriate language, which includes all the following points: (a) the words 'Important Notice' or their equivalent; (b) a statement of the superiority of breast-feeding; (c) a statement that the product should be used only on the advice of a health worker as to the need for its use and the proper method of use; (d) instructions for appropriate preparation, and a warning against the health hazards of inappropriate preparation. Neither the container nor the label should have pictures of infants, nor should they have other pictures or text which may idealize the use of infant formula. They may, however, have graphics for easy identification of the product as a breast-milk substitute and for illustrating methods of preparation. The terms 'humanized', 'maternalized' or similar terms should not be used. Inserts giving additional information about the product and its proper use, subject to the above conditions, may be included in the package or retail unit. When labels give instructions for modifying a product into infant formula, the above should apply.

9.3 Food products within the scope of this Code, marketed for infant feeding, which do not meet all the requirements of an infant formula, but which can be modified to do so, should carry on the label a warning that the unmodified product should not be the sole source of nourishment of an infant. Since sweetened condensed milk is not suitable for infant feeding, nor for use as a main ingredient of infant formula, its label should not contain purported instructions on how to modify it for that purpose.

9.4 The label of food products within the scope of this Code should also state all the following points: (a) the ingredients used; (b) the composition/analysis of the product; (c) the storage conditions required; and (d) the batch number and the date before which the product is to be consumed, taking into account the climatic and storage conditions of the country concerned.

Article 10. Quality

10.1 The quality of products is an essential element for the protection of the health of infants and therefore should be of a high recognized standard.

10.2 Food products within the scope of this Code should, when sold or otherwise distributed, meet applicable standards recommended by the Codex Alimentarius Commission and also the Codex Code of Hygienic Practice for Foods for Infants and Children.

Article 11. Implementation and monitoring

11.1 Governments should take action to give effect to the principles and aims of this Code, as appropriate to their social and legislative framework, including the adoption of national legislation, regulations or other suitable measures. For this purpose, governments should seek, when necessary, the cooperation of WHO, UNICEF and other agencies of the United Nations system. National policies and measures, including laws and regulations, which are adopted to give effect to the principles and aim of this Code should be publicly stated, and should apply on the same basis to all those involved in the manufacture and marketing of products within the scope of this Code.

11.2 Monitoring the application of this Code lies with governments acting individually, and collectively through the World Health Organization as provided in paragraphs 6 and 7 of this Article. The manufacturers and distributors of products within the scope of this Code, and appropriate nongovernmental organizations, professional groups, and consumer organizations should collaborate with governments to this end.

11.3 Independently of any other measures taken for implementation of this Code, manufacturers and distributors of products within the scope of this Code should regard themselves as responsible for monitoring their marketing practices according to the principles and aim of this Code, and for taking steps to ensure that their conduct at every level conforms to them.

11.4 Nongovernmental organizations, professional groups, institutions, and individuals concerned should have the responsibility of drawing the attention of manufacturers or distributors to activities which are incompatible with the principles and aim of this Code, so that appropriate action can be taken. The appropriate governmental authority should also be informed.

11.5 Manufacturers and primary distributors of products within the scope of this Code should apprise each member of their marketing personnel of the Code and of their responsibilities under it.

11.6 In accordance with Article 62 of the Constitution of the World Health Organization, Member States shall communicate annually to the Director-General information on action taken to give effect to the principles and aim of this Code.

11.7 The Director-General shall report in even years to the World Health Assembly on the status of implementation of the Code; and shall, on request, provide technical support to Member States preparing national legislation or regulations, or taking other appropriate measures in implementation and furtherance of the principles and aim of this Code.

Appendix 2: Resolutions of the Executive Board at its Sixty-seventh Session and of the Thirty-fourth World Health Assembly on the International Code of Marketing of Breast-milk Substitutes

Resolution EB67.R12
Draft International Code of Marketing of Breast-milk Substitutes

The Executive Board,
Having considered the report by the Director-General on the Draft International Code of Marketing of Breast-milk Substitutes;

1. *Endorses* in its entirety the Draft International Code prepared by the Director-General;
2. *Forwards* the Draft International Code to the Thirty-fourth World Health Assembly;
3. *Recommends* to the Thirty-fourth World Health Assembly the adoption of the following resolution:

28 January 1981

Resolution WHA34.22
International Code of Marketing of Breast-milk Substitutes

The Thirty-fourth World Health Assembly,
Recognizing the importance of sound infant and young child nutrition for the future health and development of the child and adult;
Recalling that breast-feeding is the only natural method of infant feeding and that it must be actively protected and promoted in all countries;
Convinced that governments of Member States have important responsibilities and a prime role to play in the protection and promotion of breast-feeding as a means of improving infant and young child health;
Aware of the direct and indirect effects of marketing practices for breast-milk substitutes on infant feeding practices;
Convinced that the protection and promotion of infant feeding, including the regulation of the marketing of breast-milk substitutes, affect infant and young child health directly and profoundly, and are a problem of direct concern to WHO;
Having considered the draft International Code of Marketing of Breast-milk Substitutes prepared by the Director-General and forwarded to it by the Executive Board;
Expressing its gratitude to the Director-General and to the Executive Director of the United Nations Children's Fund for the steps they have taken in ensuring close consultation with Member States and with all other parties concerned in the process of preparing the draft International Code;
Having considered the recommendation made thereon by the Executive Board at its sixty-seventh session;

Confirming resolution WHA33.32, including the endorsement in their entirety of the statement and recommendations made by the joint WHO/UNICEF Meeting on Infant and Young Child Feeding held from 9 to 12 October 1979;

Stressing that the adoption of and adherence to the International Code of Marketing of Breast-milk Substitutes is a minimum requirement and only one of several important actions required in order to protect healthy practices in respect of infant and young child feeding;

1. *Adopts*, in the sense of Article 23 of the Constitution, the International Code of Marketing of Breast-milk Substitutes annexed to the present resolution;
2. *Urges* all Member States:
   (1) to give full and unanimous support to the implementation of the recommendations made by the joint WHO/UNICEF Meeting on Infant and Young Child Feeding and of the provisions of the International Code in its entirety as an expression of the collective will of the membership of the World Health Organization;
   (2) to translate the International Code into national legislation, regulations or other suitable measures;
   (3) to involve all concerned social and economic sectors and all other concerned parties in the implementation of the International Code and in the observance of the provisions thereof;
   (4) to monitor the compliance with the Code;
3. *Decides* that the follow-up to and review of the implementation of this resolution shall be undertaken by regional committees, the Executive Board and the Health Assembly in the spirit of resolution WHA33.17;
4. *Requests* the FAO/WHO Codex Alimentarius Commission to give full consideration, within the framework of its operational mandate, to action it might take to improve the quality standards of infant foods, and to support and promote the implementation of the International Code;
5. *Requests* the Director-General:
   (1) to give all possible support to Member States, as and when requested, for the implementation of the International Code, and in particular in the preparation of national legislation and other measures related thereto in accordance with operative subparagraph 6(6) of resolution WHA33.32;
   (2) to use his good offices for the continued cooperation with all parties concerned in the implementation and monitoring of the International Code at country, regional and global levels;
   (3) to report to the Thirty-sixth World Health Assembly on the status of compliance with and implementation of the Code at country, regional and global levels;
   (4) based on the conclusions of the status report, to make proposals, if necessary, for revision of the text of the Code and for the measures needed for its effective application.

21 May 1981

Appendix 3: Useful addresses

Intergovernmental agencies

World Health Organization: 20 Ave. Appia, CH-1211 Geneva 27, Switzerland.

Regional offices:
Africa: PO Box No. 6, Brazzaville, Congo.
The Americas: Pan American Sanitary Bureau, 525 23rd Street NW, Washington, DC 20037, USA.
Eastern Mediterranean: PO Box 1517, Alexandria, Egypt.
Europe: 8 Scherfigsvej, DK-2100 Copenhagen O, Denmark.
South East Asia: World Health House, Indraprastha Estate, Mahatma Gandhi Road, New Delhi 110002, India.
Western Pacific: PO Box 2932, 12115 Manila, Philippines.

UNICEF: United Nations, New York, NY 10017, USA.

Regional offices:
West and Central Africa: BP 443, Abidjan, Ivory Coast.
East Africa: PO Box 44145, Nairobi, Kenya.
The Americas: Apartado Aéreo, 7555 , Bogatá, Colombia.
East Asia and Pakistan: PO Box 2–154, Bangkok, Thailand.
South Central Asia: 73 Lodi Estate, New Delhi 110003, India.
Middle East and North Africa: PO Box 811721, Amman, Jordan.
Europe: Palais des Nations, 1211 Geneva 10, Switzerland.

UN Non-Governmental Liaison Service: Palais des Nations, 1211 Geneva 10, Switzerland.

Commonwealth Secretariat: Women and Development Unit, Marlborough House, Pall Mall, London SW1, UK.

Non-governmental organizations

International Baby Food Action Network (IBFAN): CP 157, 1211 Geneva 19, Switzerland.
c/o INFACT, PO Box 300, 105, Minneapolis, MN 55405, USA.
c/o IOCU, PO Box 1045, Penang, Malaysia.

IBFAN regional centres:
South Pacific: c/o Janey Christopherson, NZCTD, PO Box 11345, Wellington, New Zealand.
South East Asia: Suseela Nair, No. 7, SS 12/3a, Subang Jaya, Selangor, Malaysia.
South Central Asia: Dr Raj Anand, 55 Kavi Apartments, Worli, Bombay 400008, India.
East and Southern Africa: Margaret Kyenkya, PO Box 34308, Nairobi, Kenya.

Caribbean: Hazel Brown, HATT, PO Box 410, Port of Spain, Trinidad.
Central America: c/o CEFEMINA, Apartado 949, San José, Costa Rica.
South America: Jeanine Anderson, Peru-Mujer, Ave. Arica 755-B, Lima 18, Peru.

Interfaith Centre on Corporate Responsibility (ICCR): Room 566, 475 Riverside Drive, New York, NY 10115, USA.

Baby Milk Action Coalition (BMAC): 34 Blinco Grove, Cambridge, UK.

War on Want: Three Castles House, 1 London Bridge St., London SE1 9SG, UK.

National Childbirth Trust: 9 Queensborough Terrace, London W2 3TB, UK.

La Lèche League International: 9616 Minneapolis Ave., Franklin Park, Illinois 60131, USA.

Industry

International Association of Infant-Food Manufacturers (IFM): 194 rue de Rivoli, F-75001 Paris, France.

Food Manufacturers' Federation (FMF): 6 Catherine St., London WC2B 5JJ, UK.

Nestlé: 1800 Vevey, Switzerland.

Cow & Gate: Trowbridge, Wiltshire, BA14 8HS, UK.

Glaxo: Clarges House, 6–12 Clarges St., London W1Y 8DH, UK.

American Home Products/Wyeth: 685 Third Ave., New York, NY 10017, USA.

Abbott/Ross: Abbott Park, N. Chicago, Ill. 60064, USA.

Bristol-Myers: 345 Park Ave., New York, NY 10022, USA.

Information

Global Village Meetings Newspapers Archive: The City University Library, Northampton Square, London EC1V OHB, UK.

Bibliography

The following works were consulted during the preparation of this book although not all are referred to directly in the text. Readers wishing to make further enquiries will find many other studies referred to in the bibliographies of some of the works listed below. The items marked with an asterisk are available in the Global Village Meetings Newspapers Archive (see Appendix 3)

Abbott/Ross Laboratories, *Code of Marketing Ethics for Developing Countries with Reference to Infant Feeding*,* Columbus, Ohio, 1977.

— *Current Concerns Regarding Infant Formula in Developing Countries*,* 1979.

— *The Abbott Laboratories Response to Third World Issues*,* 1979.

Abbott/Ross Laboratories *et al.*, *The Weakness of the Scientific Premises Underlying the WHO Code for Marketing of Breast-milk Substitutes*,* Columbus and New York, March 1981.

— *World Health Organization Development of an International Code for the Marketing of Infant Formulas: Background and Perspective*,* March 1981.

Abraham, S. C. E., *The Multinational Infant Food Industry: Towards a New Role*,* Kuala Lumpur, Malaysia, Federation of Family Planning Associations, 1978.

Abrams, E., 'The US position on infant feeding', *Nutrition Today*, July/August 1981.

Adams, S., *Roche versus Adams*, London, Jonathan Cape, 1984.

Adelman, C., 'Saving babies with a signature', *Wall Street Journal*, 28 July 1982.

Alcock, G. and Scannel, H. (eds), *Health Links*, Oxford, Third World First, 1981.

Allain, A., 'WHO Board endorse battle against bottle',* *Development Forum*, **9**, 3, April 1981.

— 'Win a battle and lose a war . . .',* in *Challenging Transnationals*, Brussels, ICDA, 1983.

— 'Not just formula',* *Health Now*, **3**, 11 May 1984.

Allain, J. P., 'Le procès des biberons accusés: les fabricants d'aliments pour bêbés', *Faim-développement*, 81–3, March 1981.

Alves, E. L. G., 'Milk supply and by-products in Brazil',* paper presented to workshop on 'Rethinking infant nutrition policies under changing socio-economic conditions', sponsored by the International Union of Nutritional Sciences, Oslo, Norway, 25 November–2 December 1980.

American Public Health Association, *Government Legislation and Policies on Pregnancy, Maternity, Infant Feeding and Marketing of Breastmilk Substitutes*,* Washington, DC, 1983.

Anand, R. K., 'The breast vs bottle controversy in India', *Indian Journal of Paediatrics*, **48**, 394, 1981.

Angst, C. L., 'The role of private enterprise in the solution of food and nutritional problems', in Santos, W. *et al.*, (eds), *Nutrition and Food Sciences: Present Knowledge and Utilisation*, Vol. I, New York, Plenum Press, 1979.

Anon, 'The infant formula debate: Third World views', *Food Policy*, **7**, 1, February 1982.

Asante, S. K. B., 'United Nations: international regulation of transnational corporations', *Journal of World Trade Law*, **13**, 1, January/February 1979.

Asiaweek, 'Battle of the Bottle',* 19 June 1981.

Baby Milk Action Coalition, *Nestlégate: Secret Memo Reveals Corporate Cover-up*,* Cambridge (UK), 1981.

Badhwar, I. and Sethi, S., 'Baby foods: breast feeding is the best formula', *India Today*, 1–15 December 1979.

Baer, E. C., 'Babies means business', *New Internationalist*, 110, April 1982.

Baer, E. C. and Margulies, L., 'An intriguing delivery: an infant formula code is born',* *Development Forum*, **9**, 6, July/August 1981.

Baer, E. C. and Winikoff, B. (eds), 'Breastfeeding: program, policy and research issues', *Studies in Family Planning*, **12**, 4, April 1981.

Baginsky, Y., 'Nestlé toe the WHO line',* *Scotsman*, 17 August 1984.

Balasubramanyam, V. N., *Multinational Enterprises and the Third World*, London, Trade Policy Research Centre, 1980.

Barnes, S., *200 Million Hungry Children*, London, Grosvenor Books, 1980.

Barness, L. A., 'Committee on Nutrition and the WHO Code of Marketing Breast-milk Substitutes',* *Pediatrics*, **68**, 3, September 1981.

Baumslag, N. *et al.*, *Breast is Best: A Bibliography on Breast Feeding and Infant Health*,* Washington, DC, US AID, 1979.

BBC Radio, 'Nature's own contraceptive',* (transcript), Tape No. TEH05/650S036, broadcast 29 April 1980.

Beaver, B. and Silvester, F., 'The gall in mother's milk: the infant formula controversy and the WHO marketing code', *Journal of Advertising*, **1**, 1, January–March 1982.

Bellini, J., 'Behind the closed doors of Nestlégate', *Vision*, May 1981.

Bemrose, C., 'The multinational charity business', *Multinational Business*, 2, 1981.

Berg, A., *The Nutrition Factor*, Washington, DC, Brookings Institute, 1973.

Bergevin, Y. *et al.*, *Do Infant Formula Samples Affect the Duration of Breast-Feeding? A Randomized Controlled Trial** (mimeo), paper presented at Annual Meeting of Ambulatory Pediatric Association, San Francisco, 30 April–1 May 1981.

— 'Do infant formula samples shorten the duration of breast-feeding?', *Lancet*, 21 May 1983, p. 1148.

Bernstein, H. (ed.), *Underdevelopment and Development*, London, Penguin, 2nd edn., 1976.

Bersier, B. C., *Memorandum of Facts*,* Vevey, Nestlé, 1979.

Biddulph, J., 'Impact of legislation restricting the availability of feeding bottles in Papua New Guinea', *Nutrition and Development*, **3**, 2, 1980.

Blum, R. *et al.* (eds), *Pharmaceuticals and Health Policy: International Perspectives on Provision and Control of Medicines*,* Penang and London, Health Action International/Social Audit, (paperback edn.), 1983.

Bonner, A., 'Improper use of baby formula not only an overseas problem',* *Washington Post*, 19 June 1981.

Borgoltz, P. A. and Latham, M. C., 'Effects of artificial feeding on nutritional status of the infant population of Lima, Peru', *American Journal of Clinical Nutrition*, **32**, 4, April 1979.

Boyer, N. A., 'WHO and a drug code',* letter to the editor, *International Herald Tribune*, 28–29 January 1984.

— *Intervention by Neil A. Boyer, Delegate of the United States in Committee A, World Health Assembly*,* (mimeo), 16 May 1984.

Brack, D. C., 'Social forces, feminism and breastfeeding', *Nursing Outlook*, **23**, 8, September 1975.

Braithwaite, J., *Corporate Crime in the Pharmaceutical Industry*, London, Routledge & Kegan Paul, 1984.

Brandt, W. *et al.*, *North-South: A Programme for Survival–Report of the Independent Commission on International Development Issues*, London, Pan, 1980.

Bristol–Myers, *Policies and Practices: Production, Labelling and Marketing of Infant Formulas*,* New York, 1978.

Brooks, P., 'Nestlé is declared "vital" by Marcos in attempt to break strike',* *Multinational Monitor*, April 1982.

Brooks, R. A., 'UNICEF, beware—dangerous shoals ahead',* *The Heritage Foundation Backgrounder*, 287, 30 August 1983.

Bulletin Today (Philippines), 'Ullmann charge probed',* 1 April 1983.
Business Abroad, 'Nestlé bolsters its outposts', 21 December 1957.
Business India, 'The dirty sales tricks of baby food giants', 18 January 1981.
Business International, 'ICC takes big step forward in drafting code of investment conduct', 2 February 1973.
—— 'What next for the international firm? A critic's view', 23 March 1973.
—— 'The international firm versus the nation state: the UN pinpoints problems and offers solutions', 17 August 1973.
—— 'United Nations report on international corporations: a guide for sensitive dealings with developing lands', 14 June 1974.
—— 'Dissent from UN report puts MNC role into perspective', 19 July 1974.
—— 'Self-regulating corporate behaviour: Caterpillar's new code of international conduct', 18 October 1974.
—— 'MNCs under fire: the challenge to chief executive officers', 6 June 1975.
—— 'How embattled MNCs can devise strategies for external affairs', 12 December 1975.
—— 'The UN's next moves on multinational firms: who wants what?', 26 March 1976.
—— 'US-based MNCs find State Dep't Committee open to their views', 16 April 1976.
—— 'Near-completed OECD Code applies to governments as well as companies', 11 June 1976.
—— 'Milk marketers in LDCs change sales techniques in response to attacks', 8 October 1976.
—— 'MNCs find a new friend—ILO', 25 February 1977.
—— 'Activist shareholders take firms to task for questionable practices', 22 April 1977.
—— 'UN continues probe into MNCs: moves to unveil transfer pricing policies', 12 August 1977.
—— 'The corporation's critics: how one influential church group operates', 10 February 1978.
—— 'Technology transfer code looms as threat to licensing freedom', 5 May 1978.
—— 'LDCs are getting expert UN assistance in dealing with MNCs', 19 May 1978.
—— 'UN study sees dulling of MNCs' bargaining edge as LDCs assert themselves', 23 June 1978.
—— 'Consumer selling in Asia: how Foremost serves up success with dairy goods', 7 July 1978.
—— 'Why MNCs fail to act before emerging issues arouse regulatory fervor', 17 November 1978.
—— 'Watching the watchdogs: a guide through the maze of codes of conduct', 29 December 1978.
—— 'Pharmaceutical industry takes varied approaches to threat to LDC sales', 9 February 1979.
—— 'Executive viewpoint', interview with Arthur Furer, chief executive officer of Nestlé, 1 June 1979.
—— 'Intracorporate pricing: UNCTAD study raps MNCs, calls for more regulation', 1 June 1979.
—— 'Advertising in LDCs: How to fit techniques to market and culture', 8 June 1979.
—— 'Questionable methodology casts doubts on findings of UN advertising study', 22 June 1979.
—— 'UN study softens position on controlling drug firms but is still critical', 6 July 1979.
—— 'What MNCs should know and why they should care about UN watchdog agency', 13 July 1979.
—— 'Multinational unions: still at incipient stage but a challenge to MNCs', 10 August 1979.

—— 'The MNC under pressure: critics rally around nutrition bugaboo', 31 August 1979.
—— 'Speaking up pays off at critical UN meeting on technology transfers', 7 September 1979.
—— 'UNCTAD rules governing anticompetitive practices quietly nears completion', 2 November 1979.
—— 'Some companies work hard to reduce the threats from codes of conduct', 21 December 1979.
—— 'MNCs are one force behind new WHO code: will strategy backfire?', 20 June 1980.
—— 'Corporate critics build steam for anti-MNC campaigns throughout Third World', 17 October 1980.
—— 'Critics of the MNC—II: key groups that examine operations in Latin America', 14 November 1980.
—— 'Consumer groups are adding muscle and becoming feistier', 21 November 1980.
—— 'Critics of the MNCs—III: churches expand strategies for prodding companies', 28 November 1980.
—— 'Critics of the MNCs—IV: several groups in the UK spell trouble for firms', 9 January 1981.
—— 'Non-activist critic helps others frame shareholder resolutions', 6 February 1981.
—— 'Ethics in the 1980s: corporations will feel pressures from within', 20 March 1981.
—— 'Firms differ sharply in how they deal with rising criticism', 3 April 1981.
—— 'Infant formula critics now planning big offensive against drug and health firms', 26 June 1981.
—— 'Chemical and drug firms may be facing restrictions on LDC marketing tactics', 5 February 1982.
—— 'Lessons from the anti-infant formula campaign', 5 February 1982.
—— 'Drug firms head off international restrictions with ideas of their own', 26 March 1982.
—— 'Nestlé tailors marketing of infant formulas to meet WHO code demands', 30 April 1982.
—— 'Even the most damned MNCs can blunt critics' charges—more on the Nestlé case', 31 December 1982.
—— 'Dealing with critics: some lessons learned in Latin America', 15 April 1983.
—— 'Consumer group plans new drive against MNC marketing', 31 August 1984.
Business Week, 'How Nestlé revives its money-losers', 7 January 1973.
California Pediatrician, 'Education grant', winter 1981, p. 9.
Carballo, M., 'Breast feeding: the natural option',* *World Health*, August/September 1979.
Carpenter, H. and Hope, R., 'The value of milk to the Jamaican economy',* letter to the editor, *Journal of Tropical Pediatrics*, **29**, August 1983, p. 196.
CBS Reports, transcript of TV documentary 'Into the mouths of babes',* broadcast 5 July 1978, New York.
CCPD/ISIS, *Report on Baby Foods Meeting*,* (mimeo), Geneva, July 1976.
Chetley, A., *The Baby Killer Scandal*,* London, War on Want, 1979.
—— 'Bottling up the formula firms',* *Development Forum*, June 1980.
—— 'Pressure increases on infant food industry',* *Outlook* (War on Want), 7, June/July 1980.
—— 'The baby milk controversy—a rejoinder',* *Food Policy*, **5**, 3, August 1980.
—— 'Marketing breast milk substitutes',* letters to the editor, *Lancet*, 2 August 1980; 18 October 1980; 29 November 1980.
—— *The Crisis in Infant Feeding*,* London, War on Want, 1981.
—— 'Breast is best: the World Health Organization's International Code of Marketing of Breast-milk Substitutes',* *IFDA Dossier*, 25, September/October 1981.
—— 'The infant formula debate: marketing code approved',* *Food Policy*, **7**, 1, February 1982.

Chetley, A., *Bangladesh: Finding the Right Prescription*,* London, War on Want, 1982.
— 'There's no substitute for mother's milk',* *World Health*, February/March 1983.
— 'Policy into practice',* *Development Forum*, June/July 1983.
— 'Working with political structures—the experience of the International Baby Food Action Network'* (mimeo), June 1983 (text of a talk delivered at Food Strategies Workshop, Lyons, 2–5 June 1983).
— 'Our campaign has just begun',* *Ideas Forum* (UNICEF), 14, 1983.
— 'Compromise on Nestlé—but the baby food campaign continues',* *Athens News*, 25 February 1984.
— 'Rule-breakers',* *Health Now*, 4, 14 May 1984.
— 'Infant feeding: US again says no',* *Health Now*, 5/6, 17 May 1984.
— 'First steps to better infant health',* *Food Policy*, **9**, 2, May 1984, p. 90.
— *A Dangerous Prescription*,* London, War on Want, 1984.
Chetley, A. and Clement, D. (eds), *Breast is Best: From Policy to Practice*,* Geneva, Minneapolis and Penang, IBFAN, 1982.
Chicago Sun-Times, 'Infant formulas lacking vital nutrient recalled', 4 March 1982.
Chorlton, P., 'Sour reception for new code on powdered milk marketing', *Guardian*, 2 August 1983.
Churchill, J. H. *et al.*, *Memorandum of Law in Opposition to Defendant's Motion for Summary Judgement*,* US District Court, Southern District of New York, 27 September 1976.
Ciocca, H. G., 'The infant formula controversy: a Nestlé view', *Journal of Contemporary Business*, **7**, 4, 1979.
City of New York, *A Local Law to Amend the Administrative Code of the City of New York, in Relation to the Provision of Safe and Adequate Nutrition for Infants by Protection of Informed Infant Feeding Choice by Mothers and Expectant Mothers** (draft), Doc. No. Int. No. 527, 14 April 1983.
Clairmonte, F. F., 'US food complexes and multinational corporations: reflections in economic predation', *Economic and Political Weekly* (India), **XV**, 41, 42 and 43, 1980.
Clark, J., *Milking Whom?–A Study of Europe's Leading Agricultural Sector and its Effect on European and Third World Food Systems*,* London, ICDA, 1979.
Clarkson, F., 'Formula for funding',* *Fairfield County Advocate*, 4 May 1983.
Clement, D., *Infant Formula Marketing in Singapore*,* Minneapolis, INFACT, November 1982.
— *Infant Milk Marketing in Indonesia*,* Minneapolis, INFACT, November 1982.
Codex Alimentarius Commission, *Recommended International Standards for Foods for Infants and Children*,* Ref. CAC/RS 72/74–1976, Rome, 1976.
Coleman, V., *The Medicine Men*, London, Arrow Books, 1977.
Commission of the European Communities, *Note: Subject: International Code of Marketing of Breast-milk Substitutes–Discussions with the Members of the Working Party on Infant Food of the Scientific Committee for Food*,* Brussels, 3 February 1982.
— *Comments by the European Economic Community on the International Code of Marketing of Breast-milk Substitutes*,* Doc. SEC(82) 316, Brussels, 25 February 1982.
— *Record of the Meeting of the Sub-group 'Dietary Foods' of the Working Party of Government Representatives on 'Foodstuffs Legislation'*,* Brussels, 24–25 March 1983, Doc. No. III/610/83–EN.
— *Working Paper: Draft Proposal for a Council Directive on the Approximation of the Laws of the Member States Relating to Infant Formulae and Follow-up Milks*,* Doc. No. III/746/83–EN, 1983.
— *Proposal for a Council Directive on the Approximation of the Laws of the Member States Relating to Infant Formulae and Follow-up Milks*,* Doc. No. COM(84) 703 final, 14 December 1984.

Constantino-David, K., 'Issues in community oragnisation',* *IFDA Dossier*, 23, May/June 1981.
Consumers' Association (UK), 'How to buy vitamins', *Which?*, January 1984.
Consumers' Association of Penang, *The Other Baby Killer*,* Penang, 1981.
Consumer Consultative Committee (of the EEC), *Preliminary CCC Opinion on the Implementation of the International Code of Marketing of Breast-milk Substitutes*,* Doc. No. CCC/32/83 final, Brussels, March 1983.
Costello, J. M. and Weinberger, M., 'Prescription drug product liability: shall patients practice medicine?', *Journal of Products Liability*, **2**, 1978.
Cottingham, J., *Bottle Babies*,* Geneva and Rome, ISIS, 1976.
Cox, D., 'A perspective on infant formula in developing countries',* (mimeo), paper presented to MIT–Harvard International Nutrition Planning Seminar, Cambridge, Mass., 12 May 1980.
— Letter to WHO and UNICEF,* Columbus, Ohio, Abbott/Ross Laboratories, 17 September 1980.
— 'The infant formula controversy: Galileo revisited'* (mimeo), paper presented at the School of Organization and Management, Yale University, 26 March 1981.
— *Statement of David O. Cox before the House Sub-committee on International Economic Policy and Trade and Sub-committee on Human Rights and International Organizations** (mimeo), Columbus, Ohio, Abbott/Ross Laboratories, 16 June 1981.
Cunningham, A. S., 'Morbidity in breast-fed and artificially-fed infants', *Journal of Pediatrics*, **5**, 726, May 1977.
— 'Breast-milk policy', letter to the editor, *World Health Forum*, **5**, 1984.
Curzon, G. and Curzon, V. (eds), *The Multinational Enterprise in a Hostile World*, London, Macmillan, 1977.
Daily Sketch (Nigeria), 'Baby food banned',* 13 November 1981.
Daily Telegraph (UK), 'US boycott of Nestlé lifted',* 27 January 1984.
Daily Times (Nigeria), 'Two brands of baby food banned from circulation',* 13 November 1981.
d'Antrim, P., 'The Nestlé product manager as demigod', in Thorelli, H. B. (ed.), *International Marketing Strategy*, Penguin, London, 1973.
Das, R., 'Impact of host government regulations on MNC operation: learning from Third World countries', *Columbia Journal of World Business*, spring 1981.
Deitch, R., 'Marketing of breast-milk substitutes in developing countries', *Lancet*, 7 March 1981.
De Moerloose, J., *Foods for infants and young children: a survey of relevant national legislation*,* Doc. HLE/FHE/81.1, Geneva, WHO, 1981.
Department of Health and Social Security (UK), *Present-Day Practice in Infant Feeding*, London, 1974.
— *Breast Feeding*, London 1978.
— *Present-Day Practice in Infant Feeding 1980*, London, 1980.
— *Press Release*: 'Code of practice on breast-milk substitutes—voluntary code of practice proposed',* London, 21 December 1982.
— *Press Release*: 'Code of practice for breast-milk substitutes',* London, 27 July 1983.
— *Health Circular*: 'Health Services Development—International Code of Marketing of Breast-milk Substitutes',* Doc. No. HC(83)13, London, July 1983.
Dogra, B., *The Milk Muddle*, New Delhi, Social Change Papers, 1980.
Dugdale, A. E., 'Breast-milk policy should be more scientific', *World Health Forum*, **4**, 1983.
Dunne, N. and McDermott, A., 'Nestlé boycott could be suspended soon',* *Financial Times*, 27 January 1984.
Ebrahim, G. J., *Breast Feeding: the Biological Option*, London, Macmillan, 1978.

Economist, 'The most international company', 12 August 1967.
Economist Development Report, 'Nestlés is nicer', March 1984.
Edwards, R. and Newens, S., *The Multinationals*, London, Liberation, 1979.
Ellestad-Sayed, J. *et al.*, 'Breast feeding protects against infection in Indian infants', *Canadian Medical Association Journal*, **120**, 1979.
Erlichman, J., 'Nestlé's £2.3bn bid in the US',* *Guardian*, 5 September 1984.
European Parliament, *Committee on Development and Cooperation: Draft Report on the Motion for a Resolution Tabled by Ms Maij-Weggen and Colleagues on the Export of Baby Food to and the sale of Baby Food in the Developing Countries (Doc. 1-688/79)*,* Doc. No. PE 74.274, Strasbourg, 19 August 1981.
— *Committee on Development and Cooperation, Draft Report on the Motion for a Resolution Tabled by Mr Collins and Others (Doc. 1-1093/81), on the International Code of Marketing of Breast-milk Substitutes*,* Doc. No. PE 81.180, Strasbourg, 21 October 1982.
— *Minutes of Proceedings of the Sitting of Monday, 11 April 1983*,* Doc. No. PE 83.818, Strasbourg.
— *Verbatim Report of Proceedings*,* Strasbourg, 11 April 1983.
Evensen, S., *Relationship Between Infant Morbidity and Breast Feeding Versus Artificial Feeding in Industrialized Countries: A Review of the Literature*,* Copenhagen, WHO Regional Office for Europe, Doc. No. ICP/NUT010/6 Rev.1, March 1983.
Fazal, A., 'Consumers in the world economy: the third force', *Multinational Monitor*, **1**, 9, October 1980.
— Brave and angry: the international consumer movement's response to TNCs',* *IFDA Dossier*, 21, January/February 1981.
Firebrace, J., *Infant Feeding in the Yemen Arab Republic*,* London, Catholic Institute for International Relations and War on Want, 1981.
— *Infant Feeding in the Middle East*,* London, IBFAN, 1983.
Flache, S., 'The infant food industry and the WHO Code', *Lancet*, 21 March 1981.
Food and Drug Letter, Editorial, 17 July 1981.
Food Manufacturers' Federation (UK), *Code of Practice for the Marketing of Infant Formulae in the United Kingdom and Schedule for a Code Monitoring Committee*,* London, July 1983.
Fookes, G. A., *The Responsibilities of the Infant Food Industry: A New Beginning*,* Zurich, ICIFI, 1979.
— *The Role of the Industry in Infant Nutrition*,* Zurich, ICIFI, 1979.
— *The Role of the Infant Food Industry*,* Zurich, ICIFI, 1979.
Furer, A., *Nestlé and Baby Foods in the Third World*,* Vevey, Nestlé, 1976.
Gaither International, Inc., *A Study of Infant Feeding in Thailand (Bangkok/Thonburi Metropolitan Area and Rural Villages of the Central Region): Summary of Findings*,* New York, 1980.
Galeano, E., *The Open Veins of Latin America*, New York, Monthly Review Press, 1971.
Geach, H., 'The baby food tragedy', *New Internationalist*, August 1973.
Geddes, J. M., 'Nestlé seeking market balance', *New York Times*, 27 March 1980.
Gelardi, R. C., 'The infant formula issue: a story in simplification and escalation', *Nutrition Today*, **16**, 5, September/October 1981.
George, S., *How the Other Half Dies*, London, Penguin, rev. edn., 1977.
— 'Nestlé Alimentana SA: the limits to public relations', *Economic and Political Weekly* (India), **XIII**, 37, 16 September 1978.
Gerrard, J. W. and Tan, L. K-T., 'Hazards of formula feeding', *Keeping Abreast, Journal of Human Nurturing*, January–March 1978.
Gillie, O. (untitled), *Good Housekeeping*, February 1984, p. 91.
— 'Nestling into life', *Sunday Times (Magazine)*, 16 December 1984.
Globe and Mail (Toronto), 'City boycotts Nestlé over baby formulas', 19 June 1981.

Goodpaster, K. E. and Matthews Jr., J. B., 'Can a corporation have a conscience?', *Harvard Business Review*, January–February 1982.

Graham, G. G., 'Comments on the World Health Organisation's "International Code" of Marketing Breast-milk Substitutes', *American Journal of Diseases in Childhood*, **135**, October 1981.

Grant, L., 'Boycott ended, Nestlé turns its attention to growth in US', *Los Angeles Times*, 5 February 1984.

Grantham-McGregor, S. M. and Back, E. H., 'Breast feeding in Kingston, Jamaica', *Archives of Diseases in Childhood*, **45**, 404, 1970.

Green, M. and Massie Jr., R. (eds), *The Big Business Reader*, New York, The Pilgrim Press, 1980.

Greiner, T., *The Promotion of Bottle Feeding by Multinational Corporations: How Advertising and the Health Professions Have Contributed*, Ithaca, NY, Cornell University International Nutrition Monograph Series, No. 2, 1975.

— *Regulation and Education: Strategy for Solving the Bottle Feeding Problem*, Ithaca, NY, Cornell University International Monograph Series, No. 4, 1977.

— 'The WHO International Code of Marketing of Breast-milk Substitutes: a time for reevaluation',* *IFDA Dossier*, 31, September/October 1982.

Greiner, T. *et al*., *The Economic Value of Breastfeeding (with results from research conducted in Ghana and the Ivory Coast*,* Ithaca, NY, Cornell University International Nutrition Monograph Series, No. 6, 1979.

Greiner, T. and Latham, M. C., 'The influence of infant food advertising on infant feeding practices in St. Vincent', *International Journal of Health Services*, **12**, 1, 1982.

Groener, D. A., 'I'm proud of my industry', *Food Monitor*, January/February 1980.

Guardian, 'Charity claims it refused $1m offer from Nestlé',* 7 February 1983.

— 'Nestlé satisfies boycott group',* 27 January 1984.

Guest, I., 'Milk and water reaction to babyfood code',* *Guardian*, 11 February 1981.

— 'Down-to-earth problems still go on shuttling',* *Guardian*, 22 April 1981.

— 'America attacks drug code plan',* *Guardian*, 21 January 1984.

— 'Which comes first—the purity of American capitalism or curing the sick?',* *Guardian*, 23 January 1984.

Gunn, R. A. *et al*., 'Bottle feeding as a risk factor for cholera in infants', *Lancet*, 6 October 1979.

Hagan, C. F., *Statement of Charles F. Hagan, American Home Products Corporation, Before the Sub-Committee of Human Rights and International Organizations and of International Economic Policy and Trade** (mimeo), New York, American Home Products/Wyeth, 16 June 1981.

— Letter to US Congress Representatives,* New York, American Home Products/ Wyeth, 10 July 1981.

Hall, B., 'Changing composition of human milk and early development of an appetite control', *Lancet*, 5 April 1979.

Halpern, J., 'The infant formula debate: spirit of cooperation needed', *Food Policy*, **7**, 1, February 1982.

Hayter, T., *The Creation of World Poverty*, London, Pluto Press (in association with Third World First), 1981.

Health Education Council (UK), *Comments from the Health Education Council on the United Kingdom's Response to the World Health Organisation's Code of Marketing of Breast-milk Substitutes*,* London, February 1982.

Health Now, 'Qatar tables baby food motion',* No. 4, 14 May 1984.

Heatley, R., *Poverty and Power*, Zed Press (in association with Returned Volunteer Action), 1979.

Heller, T., *Poor Health, Rich Profits: Multinational Drug Companies and the Third World*, London, Spokesman Books, 1977.
Helsing, E., 'Infant feeding practices in Northern Europe', *Assignment Children*, 55/56, Geneva, UNICEF, 1981.
Hencke, D., 'Britain "intends to break baby food health code"',* *Guardian*, 15 January 1983.
Hendy, R. A., *Letter to Members of the European Parliament on Behalf of the Food Manufacturers' Federation (UK)*, together with appendices, including a critique of a report which led to a resolution on the export of baby food to and the sale of baby food in the developing countries,* London, FMF, 7 October 1981.
Hilts, P. J., 'UNICEF hits Nestlé on infant formula',* *Washington Post*, 10 June 1982.
— '6–1/2 year boycott of Nestlé is ended as firm adopts baby-formula code',* *Washington Post*, 27 January 1984.
Hollie, P. G., 'Carnation: A family company's evolution',* *Wall Street Journal*, 5 September 1984.
Howell, L., 'Corporate power yields to people pressure', *Christianity and Crisis*, 16 April 1984.
IBFAN, *IBFAN News* (monthly), Minneapolis.
— *A Capsule Summary of Events at the 33rd WHA** (mimeo), Geneva, May 1980.
— *Infant Formula Promotion 1980*,* Geneva, London and Minneapolis, May 1980.
— *Baby Food Facts Foil Industry Attacks*,* Geneva and Minneapolis, September 1980.
— *Infant Formula Promotion: 100 New Violations*,* Geneva, September 1980.
— *Report of a Meeting of IBFAN** (mimeo), Geneva, 28 September 1980.
— *The Marketing of Infant Feeding Products: Proposals for Appropriate Control*,* Geneva and Minneapolis, 1980.
— *Line by Line Commentary on the WHO/UNICEF International Code of Marketing of Breast-milk Substitutes*,* Geneva, 1980.
— *Infant Formula Promotion 1980*,* Geneva, London and Minneapolis, January 1981.
— *Quotations by National Governments on WHO/UNICEF Code of Marketing and WHO Executive Board Resolution*,* Geneva, May 1981.
— *Infant Formula Promotion*,* Geneva, London and Minneapolis, 1981.
— *The Third World Speaks Out About the Dangers of Artificial Feeding*,* Geneva and Minneapolis, 1981.
— *Report of the First Congress*,* Geneva and Minneapolis, 1981.
— *Breaking the Rules*,* Geneva, Minneapolis and Penang, 1982.
— *Between the Lines: An Analysis of International Code Implementation*,* Geneva, January 1983.
— *Breaking the Rules–1982*,* Geneva, Minneapolis and Penang, January 1983.
— *Breaking the Rules 1984 in Africa*,* London, 1984.
— *Nestlé: Attempting to Rewrite History*,* London, 1984.
— 'European IBFAN groups launch Milupa campaign',* *IFDA Dossier*, September/October 1984, p. 85.
— *Breaking the Rules 1984*,* Minneapolis, December 1984.
ICCR, 'The bottle baby lawsuit', CIC Brief in the *Corporate Examiner*, April 1977.
— 'Confronting the US infant formula giants', the *Corporate Examiner*, **II**, 7–8, July–August 1982.
— *In Support of the Nestlé Boycott*, New York, February 1983.
ICDA, *The International Marketing of Breast-milk Substitutes*,* Brussels, 1983.
ICIFI, *Code of Ethics and Professional Standards for Advertising, Product Information and Advisory Services for Breast-milk Substitutes*,* Zurich, 1975 (amended 1976).
— *Constitution and Articles of Association*,* Zurich, 1975 (amended 1977, 1978 and 1979).

— *Annual Report** (covering November–September 1976), Zurich, 1976.
— 'Infant feeding in the Less Developed Countries: an industry viewpoint',* *PAG Bulletin*, **7**, 3–4, 1977.
— 'International Council of Infant Food Industries: its aims and progress', *Lancet*, 10 June 1978.
— *A Statement by the International Council of Infant Food Industries on the WHO/UNICEF Meeting on Infant and Young Child Feeding*,* Zurich, 1979.
— *Supplementary Background Paper for the Meeting on Infant and Young Child Feeding*,* Zurich, 1979.
— *The WHO/UNICEF Meeting on Infant and Young Child Feeding. Is the Industry Observing the Recommendations? The Facts*,* Zurich, August 1980.
— *National Code of Marketing in Developing Countries for Infant Formula and Other Products Intended for Use as Breast Milk Substitutes*,* Zurich, 1980.
— *Objectives, History and Activities*,* Zurich, 1980.
— *Charter for the Provision of Safe and Adequate Nutrition for Infants*,* Zurich, 1981.
— *Feeding Babies in the Third World: Industry Responds to Misinformation*,* Zurich, 1981.
— *ICIFI Observations on the Shortcomings of the Proposed WHO International Code of Marketing for Breast-milk Substitutes*,* Zurich, 1981.
— *Infant Formula Marketing in Developing Countries: Voluntary Codes and Monitoring Procedures Work*,* Zurich, 1981.
IDACE, *Industry Code of Practice for the Marketing of Breast Milk Substitutes in the EEC*,* Commission of the European Communities, Doc. No. III/1316/83–EN, revised version, 5 September 1983.
INBC, *Record of Proceedings (Mexico Meeting)*,* Toronto/New York, February 1984.
— *Record of Proceedings (Steering Committee Meeting)*,* New York, 28–29 September 1984.
INFACT, *INFACT Newsletter** (quarterly), Minneapolis.
— *INFACT Update** (monthly), Minneapolis.
— *Policy vs Practice: the Reality of Formula Promotion*,* Minneapolis, 1979.
— *Notes on the Consultation of Third Draft of WHO/UNICEF Code of Marketing** (mimeo), Minneapolis, September 1980.
— *Editorials and Opinions*,* Minneapolis, 1981.
— *Nestlé Violations of the WHO Code: Sept.-Nov. 1981*,* Minneapolis 1982.
— *Baby Bottle Disease: A Call for Action*,* Minneapolis, 1982.
— *Summary Report on the International Nestlé Boycott Meeting, 27-29 May 1983, Gex, France*,* Minneapolis, 1983.
— *An INFACT critique of NIFAC's Fourth Quarterly Report*,* Minneapolis, 1983.
International Foundation for Development Alternatives, 'Building blocks for alternative development strategies',* *IFDA Dossier*, 17, May/June 1980.
International Organization of Consumers' Unions, *Draft Code of Practice for Advertising of Infant Foods*,* The Hague, December 1971.
— *Anabolic Steroids: Availability and Marketing*, Penang, August 1983.
International Union of Food and Allied Workers' Association, *Nestlé: An IUF Survey*,* Geneva, 1970.
— *The food industry, development and trade union policy*,* Geneva, 1979.
Jacobs, B. A., 'Ross Labs finds the formula', *Industry Week*, 8 December 1980.
Janowitz, B. *et al.*, 'Breast feeding and child survival in Egypt', *Journal of Biosocial Science*, **13**, 3, July 1981.
Janus, N. and Roncagliolo, R., 'Advertising, mass media and dependency', *Development Dialogue*, 1, 1979.
Jelliffe, D. B., 'Commerciogenic malnutrition?', *Food Technology*, **25**, 55, 1971.

Jelliffe, D. B. and Jelliffe, E. F. P., 'Breast is best', *New England Journal of Medicine*, 27 October 1977.
— *Human Milk in the Modern World*, Oxford, Oxford University Press, 1978.
— 'Breast-milk policy', letter to the editor, *World Health Forum*, **5**, 1984.
— (eds), *Advances in International Maternal-Child Health*, Vol. I, Oxford, Oxford University Press, 1981; Vol. II, 1982.
Johnson, D., 'The boycott must continue', *Food Monitor*, 14, January–February 1980.
Joseph, F. G., 'Putting the bottle in its proper place',* *Development Forum*, July/August 1982.
Joseph, S. C., *Statement of Stephen C. Joseph, M.D., M.P.H., in Opposition to the US Government position on the WHO/UNICEF International Code on the Marketing of Breast-milk Substitutes** (mimeo), paper presented to meeting of American Public Health Association, Washington, DC, 18 May 1981.
— 'The anatomy of the infant formula controversy', *American Journal of Diseases in Childhood*, **135**, October 1981.
Jurt, M., '"Feldarbeit" auch für Nestlé', *Bilanz*, July 1981.
Kandel, R. F., 'A selective survey of national legislation on foods for infants and young children', *Food and Nutrition*, 7, 1, 1981.
Karanja, M., 'Kenyans "star" in bottle-baby film',* *Standard* (Kenya), 23 June 1982.
Kennedy, I., 'Now, more than ever, wealthier means healthier', *The Listener*, 13 November 1980.
Kerstiens, T., 'Are NGOs in?',* *Ideas Forum*, UNICEF, 8, 1982.
Khor, K. P., 'Value for people: the potential role of a consumer movement in the Third World',* *IFDA Dossier*, 18, July/August 1980.
Kleinman, R. L. and Senanayake, P., *Breast Feeding, Fertility and Contraception*,* International Planned Parenthood Federation, London, 1984.
Knodel, J., 'Breast feeding and population growth', *Science*, **198**, 16, December 1977.
Krause, A., 'World health agency to weigh code on baby food',* *International Herald Tribune*, 4 May 1981.
Lall, S., 'Brandt on transnational corporations, investment and the sharing of technology', *Third World Quarterly*, **II**, 4, October 1980.
— *The Multinational Corporation*, London, Macmillan, 1980.
Lambert, J. and Basford, J., 'Port Moresby infant feeding survey',* *Papua New Guinea Medical Journal*, **20**, 4, December 1977.
Lancet, 'The infant food industry', editorial, 4 September 1976.
— 'The infant food industry', editorial, 10 June 1978.
— 'Infant food industry's misgivings about next month's Geneva meeting', 22 September 1979.
— 'Code of marketing practice', 7 June 1980.
— 'Monitoring the infant formula market', 2 August 1980.
— '35th World Health Assembly' and 'Response by Nestlé to the WHO Code on Breast-milk Substitutes', 22 May 1982.
— 'Pharmaceutical policies for the Third World: whose responsibility?', editorial, 16 July 1983.
Lappe, F. M. and Collins, J., *Food First*, London, Souvenir Press, 1980.
Larsen Jr., S. A. and Homer, D. R., 'Relation of breast versus bottle feeding to hospitalization for gastroenteritis in a middle-class US population', *Journal of Pediatrics*, **92**, 1978.
Latham, M. C., *The Case Against Nestlé*,* Minneapolis, INFACT, 1978.
Lean, G., 'Doctor quits over baby food',* *The Observer*, 15 March 1981.
Ledogar, R. J., *Hungry for Profits*,* New York, IDOC, 1975.

Lefever, E. W., 'Politics and baby formula in the Third World',* *Wall Street Journal*, 14 January 1981.
LeMaire, W., 'Transnationals in the Third World', *Food Engineering International*, October 1981.
Lemaresquier, T., 'Beyond infant feeding: the case for another relationship between NGOs and the UN system',* *Development Dialogue*, 1, 1980.
Lemmers, J., 'Issue networks join forces',* *ICDA News*, March 1984.
Liebhaberg, B., *Industrial Relations and Multinational Corporations in Europe*, Farnborough, Gower (in association with the European Centre for Study and Information on Multinational Corporations), 1980.
Liotard-Vogt, P., *Address given at Nestlé Annual General Meeting** (Lausanne), Vevey, Nestlé, 22 May 1980.
Lucey, J. F., 'Does a vote of 118 to 1 mean the USA was wrong?', *Pediatrics*, **68**, 3, September 1981.
Mackenzie, D., 'From the mouths of children', *New Scientist*, 17 June 1982.
Maddison, R. E., 'Multinational companies and the EEC', *Multinational Business*, 3, 1979.
Malaysian Business, 'The $40 million a year market', May 1977.
Malcolm, A. G., *Address to the 36th World Health Assembly* (mimeo), New Zealand Ministry of Health, 4 May 1983.
Manila Times, 'Multinationals vs breastfeeding (a case of malpractice)',* 6 April 1983.
Manitoba Department of Health, 'What does the WHO "International Code of Marketing of Breastmilk Substitutes" mean for Canada?',* *Nutrition Notes for Nurses*, **VII**, 9, September 1981.
Manoff, R. K. and Cooke, T. M., 'Whose milk shall we market?', *Journal of Tropical Paediatrics*, April 1980.
Margulies, L., 'A critical essay on the role of promotion in bottle feeding',* *PAG Bulletin*, **7**, 3–4, September–December 1977.
Marin, P. C. *et al.*, 'Energy content of breast milk of poor Brazilian mothers', *Lancet*, 28 January 1984.
Marshall, E., 'Nestlé letter stops NIH talk', *Science*, **219**, 4584, 4 February 1983.
Martin, J., 'Passionate politics in a milk and water affair',* *Financial Times*, 22 May 1981.
Martin, J. and Monk, J., *Infant Feeding 1980*, London, Office of Population Censuses and Surveys, 1982.
McComas, M., *Europe's Consumer Movement: Key Issues and Corporate Responses*, Geneva, Business International, 1980.
McComas, M., Fookes, G. and Taucher, G., *The Dilemma of Third World Nutrition: Nestlé and the Role of Infant Formula*,* Vevey, Nestlé, 1983.
McCullough, T. and Gussler, J., *Factors Influencing the Decline of Breastfeeding in Developing Countries*,* Columbus, Ohio, Abbott/Ross Laboratories, 1979.
McGregor, A., '157 nations to get baby food code',* *Times*, 22 May 1981.
McKaig, D., 'Consumer-issue strategies can be shared on a worldwide basis', *International Advertiser*, July/August 1981.
May, C. D., 'The "infant formula controversy": a notorious threat to reason in matters of health', *Pediatrics*, **68**, 3, September 1981.
Medawar, C., *Insult or Injury?*, London, Social Audit, 1979.
— *The Wrong Kind of Medicine?*, London, Consumers' Association and Hodder & Stoughton, 1984.
— *Drugs and World Health*, The Hague, IOCU, 1984.
Medawar, C. and Freese, B., *Drug Diplomacy*, London, Social Audit, 1982.
Melrose, D., *The Great Health Robbery*, Oxford, Oxfam, 1981.
— *Bitter Pills*, Oxford, Oxfam, 1982.

Melville, J., 'Which ads are you sold on?', *Guardian*, 21 September 1983.
Mendelsohn, R. S., *Confessions of a Medical Heretic*, New York, Warner Books, 1980.
Middle East Health, 'Getting away with murder', April 1984.
Mintz, M., 'Infant formula maker battles boycotters by painting them Red',* *Washington Post*, 4 January 1981.
— 'Baby formula producers challenge UN agencies',* *Washington Post*, 18 March 1981.
— 'Administration urged to back strict code on breast-milk substitutes',* *Washington Post*, 3 May 1981.
Minus, P., 'The infant formula issue: other perspectives', *Christian Century*, 20–27 June 1979.
Monteiro, C., *The Harmful Activity of the Multinational Food-Producing Corporations on Child Nutrition in Brazil Through its Destructive Effects on the Organisation of the Production of Pasteurised Milk for Immediate Consumption** (mimeo), paper prepared for workshop on 'Rethinking Infant Nutrition Policies under Changing Socio-Economic Conditions', sponsored by the International Union of Nutritional Sciences, Oslo, Norway, 25 November–2 December 1980.
Mooney, P. R., *Seeds of the Earth*, London, ICDA, 1980.
— 'The Law of the Seed: another development and plant genetic resources', special issue of *Development Dialogue*, 1–2, 1983.
Movimento, 'Canned hunger: "science" to impose powdered milk' and 'Squashing the weak ones', São Paulo, Brazil, 28 January–3 February 1980.
Muller, M., 'Milk, money and marasmus', *New Scientist*, 28 February 1974.
— *The Baby Killer*,* London, War on Want, 1974 (3rd edn. 1977).
— 'Selling health—or buying favour?', *New Scientist*, 3 February 1977.
— *The Health of Nations*, London, Faber & Faber, 1982.
Munyakho, D. K., 'War on breast milk substitutes far from over',* *Sunday Nation* (Kenya), 24 May 1981.
— 'Kenya goes easy on baby milk code',* *Daily Nation* (Kenya), 25 January 1983.
Murray, J. and Murray, A., 'Breast milk and weights of Nigerian mothers and their infants', *American Journal of Clinical Nutrition*, **32**, 4, April 1977.
National Childbirth Trust (UK), *Breastfeeding Promotion Group's Commentary on FMF Code*,* London, February 1982.
Nelson, D., 'Nations slow to adopt baby food code', *Commonwealth*, **25**, 6, June 1983, p. 211.
Nestlé, *This is Your Company*,* Vevey, 1946.
— *Annual Report*(s),* Vevey, 1968–83.
— *Nestlé in the Developing Countries*,* Vevey, 1975.
— *Nestlé and Infant Formula: Facts and Fallacies*,* Vevey, 1979.
— *Infant and Young Child Feeding in Three Developing Countries*,* Vevey, November 1981.
— *Marketing Infant Formula in the Third World: How Nestlé Has Followed the WHO/UNICEF October 1979 Recommendations*,* Vevey, 1981.
— *The Story Behind the Issue: Infant Feeding in Developing Countries*,* Washington, DC, 1981.
— *Zimbabwe: Summary of Main Criticisms Aimed at Food Specialities*,* Vevey, February 1982.
— *WHO International Code of Marketing of Breast-milk Substitutes: Instructions to All Companies of the Nestlé Group and to Agents and Distributors Who Market Infant Formula under Trade Marks Owned by the Nestlé Group*,* Vevey, March 1982 (revised October 1982).
— *Nestlé and the WHO Code*,* Vevey, April 1983.
— *Nestlé News*, monthly from November 1983, Washington, DC.

Nestlé Infant Formula Audit Commission, *Charter, Articles of Incorporation and By-laws and Procedures*,* Washington, DC, 1982.

— *Quarterly Reports*,* First, October 1982; Second, December 1982; Third, March 1983; Fourth, June 1983; Fifth, September 1983; Sixth, December 1983; Seventh, March 1984; Eighth, June 1984; Ninth, September 1984, Washington, DC.

— *Transcript of proceedings of Press Conference with Edmund S. Muskie and Dr Carl Angst*,* Washington, DC, 14 October 1982; also 21 April 1983.

New Internationalist, 'Stop the babymilk pushers', special issue, No. 110, April 1982.

Newland, K., *Infant Mortality and the Health of Societies*,* Washington, DC, Worldwatch Institute, 1981.

New Scientist, 'Milk lobby upsets doctors', 19 February 1981.

— 'Substitute for breast milk escapes curbs', 20 January 1983.

— 'Babymilk code fails to calm critics', 11 August 1983.

Newton, N., 'The importance of prenatal education', *Child and Family*, winter issue, 1969.

New York Times, '300,000 cans of infant formula recalled over fear of illness',* 10 November 1979.

Nicely, S., 'Boycott of Nestlé is called victory',* *Kansas City Times*, 19 March 1983.

Nicholls, J., 'Pressure groups still urge tighter controls on baby food promotion', *Medical News*, 13 January 1983.

Nickel, H., 'The corporation haters', *Fortune*, 16 June 1980.

Ninomiya, K., 'Foreign firms' strategies in Japan: case studies of Nestlé Japan and Coca Cola (Japan)', *Oriental Economist*, November 1980.

Northrup, H. R. and Rowan, R. C., *Multinational Collective Bargaining Attempts*, University of Pennsylvania, 1979.

Noyce, J. L., *Marketing Infant Formula: A Bibliography*,* London, War on Want, 1982.

Nursing Times (UK), 'HVA slams voluntary code on baby milk', 10 August 1983.

Nutrition Committee of the Canadian Paediatric Society and Committee on Nutrition of the American Academy of Pediatrics, 'Breast-feeding: a commentary in celebration of the International Year of the Child, 1979', *Pediatrics*, 4, **62**, October 1978.

Nutrition Foundation of the Philippines, editorial: 'On the breast feeding campaign', *Bulletin of the Nutrition Foundation of the Philippines*, **20**, 2, March/April 1980.

Ogier, B., *Marketing Aspects of Developing Export Products* (mimeo), paper presented to seminar on the 'Importance of Adding Value to New Zealand's Food Exports', at Massey University, 15 February 1980.

Ogra, P. L. and Greene, H. L., 'Human milk and breast feeding: an update on the state of the art', *Pediatric Research*, 266–71, 16, 1982.

Owles, D., 'Pharmaceutical manufacturers' duty to warn of propensity to harm', *Journal of Products Liability*, **3**, 1979.

Pagan, R., 'Carrying the fight to the critics of multinational capitalism: think and act politically', *Vital Speeches of the Day*, **XLVIII**, 19, 15 July 1982.

— 'Resolving the Nestlé boycott: 1977–1984', *IPRA Review*, August 1984, pp. 21–6.

Pausenberger, E., 'The conduct of international enterprises in developing countries: an empirical review of some prejudices', *Intereconomics*, November/December 1980.

Pediatric Research, 'Promotion of breast feeding: recommendations of the Councils of the Society for Pediatric Research (SPR) and American Pediatric Society (APS), and of the American Academy of Pediatrics (AAP)', 264–5, 16, 1982.

Pergamon Press, 'Pharmaceuticals and health in the Third World', special issue, *World Development*, **11**, 3, March 1983.

Pfaff, C., 'New threats to multinationals', *Advertising Age/Europe*, December 1981, p. 3.

Pfizer, *Pfizer and Third World Health Issues*, New York, 1983.

Pharmaceutical Manufacturers Association (US), 'PMA President Engman on the "Third World"', *Quotes of Note*, 28 September 1981.
Philippine News, 'Liv Ullmann was "shocked" for being censored',* 6–12 April 1983.
Plank, S. J. and Milanesi, M. L., 'Infant feeding and infant mortality in rural Chile',* *Bulletin of the World Health Organization*, 48, 1973.
Population Reports, 'Breast-feeding, fertility and family planning', Series J, No. 24, Baltimore, MD, Johns Hopkins University, November/December 1981.
Post, J. E. and Baer, E., 'Analysing complex policy problems: the social performance of the international infant formula industry', in Preston, L. (ed.), *Research in Corporate Social Performance Policy*, Vol. 2, Greenwich, Conn., Jai Press, 1979.
— 'Demarketing infant formula: consumer products in the developing world', *Journal of Contemporary Business*, 7, 4, 1979.
— 'The International Code of Marketing for Breast-milk Substitutes: consensus, compromise and conflict in the infant formula controversy', *International Commission of Jurists' Review*, 25, December 1980.
Printers' Ink (US), 'Powdered milk to African tribesmen: Borden's world-wide marketing mission', 13 June 1958.
Privolos, B., 'Faced with a TV advertising ban in Indonesia', *International Management*, January 1985, pp. 28–33.
Protein–Calorie Advisory Group (of the UN), 'Promotion of special foods (infant formula and processed protein foods) for vulnerable groups',* *PAG Statement No. 23*, New York, 18 July 1972, revised 28 November 1983.
Radice, H. (ed.), *International Firms and Modern Imperialism*, London, Penguin, 1975.
Ram, N., 'Code on infant formulae: US decision revolting',* *Hindu* (Madras, India), 21 May 1981.
Rangan, R. K., 'Austere paediatric meet', *Times of India*, 23 June 1983.
Raphael, D. (ed.), *Breastfeeding and Food Policy in a Hungry World*, New York, Academic Press, 1979.
Ress, P. E., '"Kangaroo" treatment saves underweight babies',* *UNICEF News Feature*, Geneva, May 1984.
Richardson, J. L., 'Review of international legislation establishing nursing breaks', *Journal of Tropical Paediatrics*, October 1975.
Ricketts, B. D., 'Marketing breast milk substitutes', letter to the editor, *Lancet*, 15 November 1980.
Roncagliolo, R. and Janus, N., 'Advertising and the democratisation of communications', *Development Dialogue*, 2, 1981.
Rothwell, N., 'Nestlé makes a defence in storm over baby formula', *Australian*, 13 December 1982.
Rowland, M. G. M. *et al*., 'Bacterial contamination in traditional Gambian weaning foods', *Lancet*, 21 January 1978.
Salisbury, L. and Blackwell, A. G., *An Administrative Petition to the United States Food and Drug Administration and Department of Health and Human Services*,* San Francisco, Public Advocates Inc., 1981.
Sampson, A., *The Sovereign State*, London, Coronet, 1974.
— *The Seven Sisters*, New York, Bantam, 1976.
Saunders, E. W., 'Marketing breast milk substitutes', letters to the editor, *Lancet*, 5 July 1980 and 20 September 1980.
Sauvant, K. P. and Lavipour, F. G. (eds), *Controlling Multinational Enterprises: Problems, Strategies, Counter-strategies*, Boulder, Colorado, Westview Press, 1976.
Schoonmaker, M. E., 'Baby formula: a peril in the US?',* *Sunday Record* (New Jersey), 28 June 1981.

Schwamm, H., 'The OECD guidelines for multinational enterprises', *Journal of World Trade Law*, **12**, 4, July/August 1978.

Schwartz, H. 'A drug-code warm-up', *Pharmaceutical Executive*, August 1981.

Scott, P. T., 'Innocents abroad: infant food technology at the law's frontier', *Virginia Journal of International Law*, **20**, 3, 1980.

Scott-Atkinson, D., 'The Nestlé caper is a PR nightmare', *Marketing* (Canada), 12 October 1981.

Scrimshaw, N. S., 'Code not cure', *Nutrition Today*, July/August 1981.

Seabrook, J., 'The great consumer swindle', *Guardian*, 21 August 1982.

Seitz, R. H., 'Infant formula promotion in the Philippines', *Far East Health*, July 1981.

— *Breast Feeding and Infant Formula Promotion: Singapore** (mimeo), Minneapolis, INFACT, February 1982.

— *Breast Feeding and Infant Formula Promotion: Thailand** (mimeo), Minneapolis, INFACT, February 1982.

— *Breast Feeding and Infant Formula Promotion: Philippines** (mimeo), Minneapolis, INFACT, March 1982.

Sexual Medicine Today, 'The neglected contraceptive', May 1978.

Short, K., *The World Health Organisation Code of Marketing of Breast-milk Substitutes: Status of Compliance with and Implementation of the Code in Australia*,* Australian Consumers' Association, 1983.

Short, R. V., 'The biological basis for the contraceptive effects of breast feeding',* unpublished.

— 'Breast feeding', *Scientific American*, **250**, 4, April 1984.

Shuster, S., 'Under pressure', *World Medicine*, 6 March 1982.

Silverman, M., *The Drugging of the Americas*, Berkeley, University of California Press, 1976.

Silverman, M. and Lee, P., *Pills, Profits and Politics*, Berkeley, University of California Press, 1974.

Silverman, M., Lee, P. and Lydecker, M., *Prescriptions for Death: The Drugging of the Third World*, Berkeley, University of California Press, 1982.

Sjolin, S., 'Should paediatricians support infants and/or the baby food industry?', *Lancet*, 14 March 1981.

Solagral and UN Non-Governmental Liaison Service, *Report on the Food Strategy Workshop*,* Paris/Geneva, 1983.

Sommerville, F. P. L., 'Support decreases for Nestlé boycott',* *Sun* (Baltimore), 25 January 1984.

Sousa, P. L. R. *et al.*, 'Patterns of weaning in South Brazil', *Journal of Tropical Paediatrics*, August 1975.

Sparks, J. A., 'Anatomy of a boycott', *Food Policy*, **5**, 3, August 1980.

Speck, K., 'Nestlé-Chef: "Wir fühlen uns nicht eingeschrankt"', *Tages-Anzeiger*, 19 October 1979.

Standard (Kenya), 'Vital ethics of infant care',* 18 April 1981.

Starfield, B., 'The World Health Organization Code of Marketing of Breast-milk Substitutes', *Pediatrics*, **68**, 3, September 1981.

Steady, F. C., 'Infant feeding in developing countries: combatting the multinationals imperative', *Journal of Tropical Paediatrics*, **27**, August 1981.

Sterngold, J., 'Nestlé planning to pay $3 billion to acquire Carnation company',* *Wall Street Journal*, 5 September 1984.

Stokes, G., 'Nestlé's good "fortune"', *Village Voice*, 28 January–3 February 1981.

Sun (Baltimore), 'Nestlé will obey code on formula',* 17 March 1982.

Swiss Information Groups for Development Policy, *Does Nestlé Kill Babies?** (mimeo), 1975.

—— *First Hearing of Witnesses in the Nestlé Case*,* information for the press, No. 3 (mimeo), February 1976.

—— *Nestlé's Arguments** (mimeo), May 1976.

—— *Nestlé Case: The Third Hearing Begins on June 22, 1976*,* information for the press, No. 4 (mimeo), May 1976.

—— *Nestlé Case Will Not Continue–But Dispute Goes On*,* information for the press, No. 5 (mimeo), December 1976.

Tanner, J., 'Baby milk campaigners alter their sights',* *Gemini News Service*, November 1984.

—— 'Milk marketing continues', *New Statesman*, 23 November 1984, pp. 21–2.

Taylor, D., *Medicines, Health and the Poor World*, London, Office of Health Economics, 1982.

Thomas, H., 'Money makes a bad master', *Guardian*, 5 June 1984.

Thordahl, M. C., 'National and multinational opportunities for the encouragement of innovation', *Food Technology*, January 1982.

Tisdall, P., 'Medicinal maelstrom', *Marketing* (UK), 12 April 1984.

Toffler, A., *The Third Wave*, London, Pan Books, 1980.

—— 'The rushing tidal wave of change', *Sunday Times*, 17 June 1984.

Tudge, C., *The Famine Business*, London, Faber & Faber, 1977.

Tugendhat, C., *The Multinationals*, London, Penguin, 1973.

United Nations, *Towards a World Economy that Works*, New York, UN Department of Public Information, 1980.

—— *Consolidated List of Products Whose Consumption and/or Sale Have Been Banned, Withdrawn, Severely Restricted or Not Approved by Governments* (first issue revised), Doc. DIESA/WP/1, New York, July 1984.

UN Centre on Transnational Corporations, *Transnational Corporations in Advertising*, Doc. ST/CTC/8, New York, 1979.

—— *Transnational Corporations and the Pharmaceutical Industry*, Doc. ST/CTC/9, New York, 1979.

—— *Measures Strengthening the Negotiating Capacity of Governments in their Relations with Transnational Corporations: Technology Transfer through Transnational Corporations*, Doc. ST/CTC/11, New York, 1979.

—— *Transnational Corporations in Food and Beverage Processing*, Doc. ST/CTC/19, New York, 1981.

UN Commission on Transnational Corporations, *The Social and Cultural Impact of Transnational Corporations: Report of the Secretariat*, Doc. E/C.10/86, 22 June 1981.

—— *Role of the Information System on Transnational Corporations Regarding the Exchange of Information on Banned Hazardous Chemicals and Unsafe Pharmaceuticals: Report Prepared by the Secretariat*, Doc. E/C.10/90, 20 July 1981.

UNICEF, 'Breast-feeding and health', special issue, *Assignment Children*, 55/56, 1981.

—— *The State of the World's Children 1982-83*, Oxford, Oxford University Press, April 1983.

—— *Transcript of Press Briefing by Liv Ullmann*,* London, 16 June 1983.

—— *The State of the World's Children 1984*, Oxford, Oxford University Press, 1984.

—— 'Marketing survival', special issue, *Ideas Forum*, 19, 1984/5.

—— *The State of the World's Children 1985*, Oxford, Oxford University Press, 1984.

—— United Nations University, 'Consultative Group on Maternal and Young Child Nutrition', *Food and Nutrition Bulletin*, **1**, 3, May 1979.

—— 'Breast feeding versus bottle feeding in developing countries: a brief for policy makers', *Food and Nutrition Bulletin*, **3**, 2, April 1981.

USA Today, 'Ullmann claims Nestlé kept her off TV show',* 23 March 1983.

US Congress, *Marketing and Promotion of Infant Formula in Developing Countries: Hearing*

Before the Sub-committee on International Economic Policy and Trade of the Committee on Foreign Affairs of the House of Representatives, Washington, DC, US Government Printing Office, 1980.

— *Congressional Record*, No. 91, 16 June 1981; No. 92, 17 June 1981; No. 93, 18 June 1981.

US Senate, *Marketing and Promotion of Infant Formula in the Developing Countries: Report of the US Senate Health and Scientific Research Sub-Committee*, Washington, DC, US Government Printing Office, 1978.

— *Hearings Before the Committee on Foreign Affairs on Nomination of Ernest W. Lefever, to be Assistant Secretary of State for Human Rights and Humanitarian Affairs*, Washington, DC, US Government Printing Office, 1981.

Valyasevi, A. and Baker, J. (eds), *Proceedings of the Workshop on Breastfeeding and Supplementary Foods*, Bangkok, Thailand, 1979.

van der Velden, F., 'Challenging multinational corporations: case of Bristol Myers company in Netherlands', *Mainstream*, 4 August 1984, pp. 20–4.

Villar, J. and Belizan, J. M., 'Breastfeeding in developing countries', *Lancet*, 15 September 1981.

Vitale, E. *et al.*, 'Marketing a socially significant food product in a developing country: the case of a weaning food in Morocco',* unpublished.

Waldholz, M., 'Pill promoters: marketing often is the key to success of prescription drugs', *Wall Street Journal*, 28 December 1981.

War on Want, *Fruits of Development*, London, 1980.

— *Breast or Bottle? Factors Influencing the Choice of Infant Feeding in the UK*,* London, 1982.

— *The International Code of Marketing of Breast-milk Substitutes: Is It Being Followed in Europe?*,* London, October 1982.

— *Commentary on and Suggestions for Improvement of 'Draft Code of Practice for the Marketing of Infant Formulae in the United Kingdom and Schedule for a Code Monitoring Committee'*,* London, January 1983.

— *Commentary on and Suggestions for Improvement of 'Code of Practice for the EEC for the Marketing of Breastmilk Substitutes'*,* London, February 1983.

— *Baby Milk Companies Produce Worthless Marketing Code*,* London, August 1983.

— *Baby Foods Campaign Update*,* London, February 1984.

Warshaw, A., 'Breast milk code "not complied with in 1982"',* *Bangkok Post*, 24 January 1983.

Watson, R., 'Prising secrets from multinationals', *Guardian*, 16 June 1980.

Weatherby, V., 'Baby's milk: soft sell success', *Marketing* (UK), 14 May 1980.

Weitz, M., *Health Shock*, London, Hamlyn Paperbacks, 1982.

Wennen-van der May, C. A. M., 'The decline of breast feeding in Nigeria', *Tropical and Geographical Medicine*, 21, 1969.

Werner, D., *Where There Is No Doctor*, Palo Alto, Calif., Hesperian Foundation, 1977.

— 'The accountability of community health workers: to whom and for what', in *Report on Conference on Technology Transfer to the Third World*, Dhaka, Bangladesh, Gonoshasthaya Pharmaceuticals Ltd., 1982.

Willat, N., 'How Nestlé adapts products to its markets', *Business Abroad*, June 1970.

Willetts, P. (ed.), *Pressure Groups in the Global System*, London, Frances Pinter, 1982.

Wilson, M., 'A problem of ideology and social relations: bottle babies and managed myths', *Science for the People*, January–February 1981.

Winikoff, B. and Baer, E. C., *The Obstetricians' Opportunity: Translating 'Breast is Best' from Theory into Practice*,* New York, Population Council, 1980.

World Health Organization, *Joint WHO/UNICEF Meeting on Infant and Young Child Feeding: Statement and Recommendations*,* Geneva, 1979.

— *Provisional Summary Record of the Sixth Meeting of Committee A, 33rd World Health Assembly*, Doc. No. A33/A/SR/6, 16 May 1980; also for seventh meeting, Doc. A33/A/SR/7, 16 May 1980; and eighth meeting, Doc. A33/A/SR/8, 17 May 1980.

— *Infant and Young Child Feeding*, Doc. No. EB67/20, 10 December 1980.

— *Report of the Standing Committee on Non-governmental Organisations*, Doc. EB67/42, 24 January 1981.

— *Infant and Young Child Feeding: Progress Report by the Director-General*, Doc. A34/7, 19 March 1981.

— *Communication from the Delegation of the United Kingdom*, Doc. A34/INF.DOC./10, 14 May 1981.

— *Draft International Code of Marketing of Breast-milk Substitutes*, Doc. A34/INF.DOC./11 and Doc. A34/INF.DOC./12, 20 May 1981.

— *International Council of Infant Food Industries*, Doc. EB69/NGO/2, October 1981.

— *Contemporary Patterns of Breastfeeding: Report on the WHO Collaborative Study on Breast-feeding*, 1981.

— *Handbook of Resolutions and Decisions of the World Health Assembly and Executive Board*, Vol. II, 4th edn., 1981.

— *Infant and Young Child Feeding: Current Issues*, 1981.

— *International Code of Marketing of Breast-milk Substitutes*, 1981.

— *Thirty-Fourth WHA, Summary Records of Committees*, Doc. WHA34/1981/REC/3, 1981.

— *Infant and Young Child Feeding: Progress Report by the Director-General*, Doc. A35/8, 30 March 1982.

— *International Code of Marketing of Breast-milk Substitutes*, Resolution WHA35.26, 14 May 1982.

— *Basic Documents*, 32nd edn., 1982.

— *Director-General's Report*, Doc. A36/7, May 1983.

— *Women and Breast-feeding*, 1983.

— *Director-General's Report*, Doc. A37/6, May 1984.

— *Infant and Young Child Nutrition*, Resolution WHA37.30, May 1984.

WHO/UNICEF, *Joint WHO/UNICEF Meeting on Infant and Young Child Feeding: Background Paper and Themes for Discussion*,* Doc. FHE/ICF/79.3, Geneva, 1979.

— *Infant and Young Child Feeding: Notes on the International Code of Marketing of Breast-milk Substitutes*, Geneva, 1982.

— *Comparative Statements of the International Code, the Nestlé Instructions and the WHO/UNICEF notes of Clarification*, New York and Geneva, August 1982.

Worral, J., 'Baby milk curbed in Kenya',* *Sunday Times*, 26 April 1981.

— 'Baby food code welcomed',* *Financial Times*, 27 April 1981.

Wurst, J. (ed.), *Moral Pressure for Liberation*, Antwerp, Non-Violent Alternatives and Peace Action News, 1981.

Wyeth, *Wyeth Policies for Marketing of Infant Formula Products in Developing Countries*,* Philadelphia, 1978.

— *Wyeth Guidelines for Sampling of Infant Formula Products*,* 1979.

— *Policy of Wyeth Laboratories and Its Worldwide Affiliates with Regard to WHO International Code of Marketing of Breast-milk Substitutes*,* New York, 1982.

Young, R., 'Food firms upset by UN stand on baby feeding',* *The Times*, 6 March 1980.

Zimbabwe Ministry of Health, *Baby Feeding: Behind and Towards a Health Model for Zimbabwe*, Harare, 1982.

Index